HOW TO BUILD A UNIVERSE

FROM THE BIG BANG TO THE END OF THE UNIVERSE

An Hachette UK Company
www.hachette.co.uk

First published in Great Britain in 2015 by
Philip's, a division of Octopus Publishing Group Ltd
Endeavour House
189 Shaftesbury Avenue
London
WC2H 8JY
www.octopusbooks.co.uk

ISBN 978-1-844-03809-1

A CIP catalogue record for this book is available from the British Library.

Printed and bound in China

10 9 8 7 6 5 4 3 2 1

HOW TO BUILD A UNIVERSE

FROM THE BIG BANG TO THE END OF THE UNIVERSE

BEN GILLILAND

PHILIP'S

CONTENTS

INTRODUCTION

○○○

THE 'MIRACLE' OF YOU

When humankind first pondered its existence, it did so in a hostile world. Living as small nomadic groups of hunter-gatherers, early humans had no control over their destiny so they sought it by imagining that their fates were in the hands of gods. After all, nothing pierces the gloom of a short, hard life more effectively than the hope of a good miracle. Then science came along and, through the gathering of evidence and the testing of ideas, uncovered the natural laws and mechanisms that govern the cosmos. Even the miraculous could be explained with the correct application of critical thinking, evidence and experimentation. As science was dispelling the superstitious miracles, it uncovered the greatest miracle of all: the miracle of you.

Your journey began some 13.8 billion years ago in a place before space and time, on a day without a yesterday. Somewhere in the middle of nowhere, all the future potential of the Universe was bound together in an area smaller than the smallest particle. Then (for reasons still unknown) all this potential was released in a colossal 'WHOOSH' and the Universe was born. At first, a roiling soup of super-heated plasma, the Universe expanded and cooled and, as it did so, the first particles coalesced from the soup. All those particles were created in two varieties – matter and, its opposite, antimatter. Had matter and antimatter been made in equal measure, the Universe would have ended there and then – in a chain reaction of mutually assured destruction. But, for reasons we still don't understand, matter ever so slightly outnumbered antimatter and the Universe (and the potential you) continued to exist.

But your existence still wasn't a foregone conclusion. As the Universe expanded, matter spread out. Had it done so evenly (like water filling a bucket), it might have remained that way forever. Luckily, the expanding Universe wasn't perfectly even and, in pockets where stuff was just a little denser, gravity went to work. It pulled matter together to create clouds of gas, which collapsed – creating enough heat and pressure to kickstart the nuclear fusion reactions that powered the first stars, and squeezed atoms together to create the heavy chemical elements that you would be built from.

All of those chemical goodies were of no use locked away in the hearts of the stars. Fortunately, those early stars were truly massive and a massive star is a short-lived star, so, after cooking up those heavy elements, they exploded as supernovae – peppering the cosmos with their fertile seed. If the laws of physics had been slightly different, those stars might not have been massive enough to go 'KABOOM' and your chemical ingredients would have remained half-baked and locked away for eternity in the bowels of a cooling lump of carbon. After a few billion years, several cycles of nuclear fusion, galaxies forming and the Universe still existing, one region of a galaxy called the Milky Way was ready to witness the next miracle.

About 4.5 billion years ago, around an unremarkable star, a planet coalesced from a swirling disk of dust and ice. It wasn't much to look at – just a searingly hot ball of molten rock and sinking metals – but it had formed at a near-perfect distance from its star. It wasn't so close that it would remain oven-hot forever and not so distant that it would become a large, novelty ice cube. Life would stand a pretty good chance on a planet like that, but it would take one more miracle for that to occur.

That miracle arrived in the form of a Mars-sized planet, which smashed into our infant planet throwing a vast mass of rocky material into space. This formed our Moon. The impact that created the Moon also knocked the Earth sideways on its axis, which meant that the Sun's energy wasn't focused on a single region, and the gravitational presence of the Moon stopped the Earth from wobbling erratically on that axis. This stabilized the Earth's climate and prevented violent (potentially life-extinguishing) climactic swings. The Moon's creation had turned the Earth into a perfect nursery for life. But it wasn't done yet. The Moon's gravity tugged on the planet's oceans, and so began the tides that daily massage the world's coastlines today. It may have been this very tidal action of repeatedly (and regularly) exposing and then submerging the coast that actually caused life to evolve in the first place.

Here's one final miracle for you... Whatever the mechanism that caused their evolution, among those first, single-celled life forms was your ancestor. For you to be sitting here reading this today, there has been an unbroken chain of existence between you and that tiny, floating forebear. For 3.8 billion years, every one of your ancestors survived long enough to pass their genetic material on to the next generation. Just think how unlikely that is. Over almost 4 billion years of mass extinctions, predation, disease, social upheaval, war and famine, there is an unbroken chain of life that leads to you. Now that's what I call a miracle.

IN THIS BOOK

In this book we'll chart how energy became matter and how a set of physical laws guided the interactions that allowed matter to build the stars, galaxies and you. And we'll chart some of the scientific discoveries and breakthroughs that have helped us understand how to build a Universe.

HOW WE DISCOVERED THE BIG BANG
(AND LEARNED HOW TO MEASURE THE UNIVERSE)

In which we chart the chain of events that led us to believe that the Universe, rather than being static and eternal, had a moment of 'birth' and has been growing ever since

The notion that the Universe was born in a 'Big Bang', or indeed anything else, is a relatively new concept. In fact even the term 'Big Bang' was coined by someone who didn't believe in it and intended it as a disparaging put down. Today, however, Big Bang theory is one of the most successful ideas in science, but how did we get there?

From the age of the ancient Greeks to the Scientific Revolution nearly 2,000 years later, it was believed that everything in the Universe was entrapped within a series of celestial spheres that encased the Earth, which was (of course) the central pivot around which the rest of existence rotated. These celestial spheres were the Solar System, which was thought to be the full extent of the Universe.

Science moved on a bit by the 16th and 17th centuries, with the likes of the astronomer Nicolaus Copernicus and the famous Italian polymath Galileo Galilei, who used reasoning, mathematics and observation to prove that the Earth and the rest of the planets orbit the Sun.

A crucial innovation at that time was the invention of the telescope. Originally little more than an amusing curiosity, the telescope was introduced to astronomy in 1609 by Galileo Galilei and the not-so famous English polymath, Thomas Harriot (who used his telescope to sketch the Moon four months before Galileo's celebrated observations, and who is sometimes credited with introducing the potato to England).

The telescope helped increase the size at which we could view the Universe. Galileo's observations of the strange milky band that crossed the night sky revealed that it was made up of stars – the Universe had now increased in size to include the Milky Way.

BEYOND THE SOLAR SYSTEM

After a few decades of being used to peer at planets, moons and comets, the telescope was next put to use to seek out objects beyond our home galaxy. In the late 1700s,

Big Bang   Particles form   CMB   Dark ages (first dark matter structures)   First stars and active galaxies

13.82bn years ago   377,000 years after Big Bang   200 million years

STAR LIGHT, STAR (REALLY, REALLY) BRIGHT

Redshifted galaxies and stars (see page 14) are good evidence that the Universe is expanding from a point of birth, but if you don't have access to expensive telescopes, is there any way to reach the same conclusion from the comfort of your own back garden?

Luckily there is one easy way to show that the Universe can't possibly be infinite and unchanging – just look up at the sky on a cloudless night (unless you live beneath a floodlight) and you will see a mostly black sky peppered with stars. But, if the Universe is static and infinite, all the stars ...

... from here ...   and here ...   and here ...   and here...   ...would be visible from here, and the night sky would be as bright as the Sun.

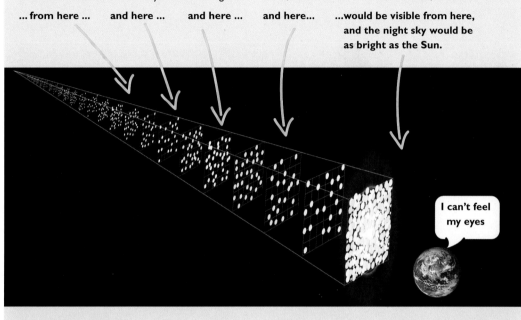

I can't feel my eyes

If the Universe is infinite and unchanging, it would contain an infinite number of stars that would all be visible from Earth.

In a Universe of infinite age, light from even the most distant stars would have had an infinite amount of time to reach us and, if the Universe is static, the light from those stars would arrive unchanged (not stretched into different parts of the spectrum).

So, in an infinite Universe, a star would be visible at every point and the night sky would be as bright as the Sun. Since it is pretty much accepted that you can't get a suntan at night, it's quite clear that the Universe must be expanding.

a Frenchman, Charles Messier – who was trying his best to discover new comets (he discovered 13 in his lifetime) – kept stumbling across strange fuzzy objects in the heavens, which he would at first mistake for comets. To avoid this confusion, Messier compiled a

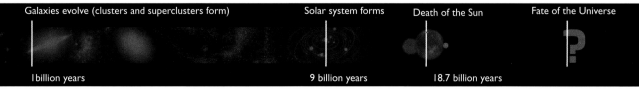

Galaxies evolve (clusters and superclusters form)   Solar system forms   Death of the Sun   Fate of the Universe

1 billion years   9 billion years   18.7 billion years

catalogue of the strange nebulous glows. By the time he died, oblivious to what he had been cataloguing, he had charted the locations of 103 of these white smudges. For the next two centuries, the identity of his 'Messier objects' remained a mystery.

By the 19th century there were two schools of thought as to the identity of the nebulae. One, championed in the previous century by the great astronomer William Herschel, was that they were 'island universes' located beyond our own Milky Way. The other, more popular, idea was that they were little more than clouds of gas floating within (or just outside of) the Milky Way.

In the 1860s, British astronomer William Huggins borrowed a trick from the field of chemistry, which moved things on a bit: spectroscopy. A spectroscope is an instrument that splits light into its component colours like raindrops split sunlight to create a rainbow – spreading out the spectrum of light into its different wavelengths. Hidden in the rainbow is a series of bright or dark lines (called emission and absorption lines) that are caused by chemical elements in the object the light is coming from. These lines act like a sort of chemical barcode that allows you to identify the element that created it.

The 'Great Forty-Foot telescope' was built in Slough, England, by William Herschel and his astronomer wife, Caroline Herschel. Completed in 1789, it was the largest telescope in the world for 50 years.

Using spectroscopy, Huggins was able to identify the elements that make up the Sun and compare the Sun's barcode with that of other stars. He found that starlight contained pretty much the same spectral barcode as the Sun – meaning that distant stars were made of the same mixture of chemical elements as the star on our doorstep.

Huggins then turned his spectroscope to Messier's nebulae and, beginning in 1864, he examined the spectra of about 70. Around a third of the clouds didn't exhibit the spectral patterns of stars but instead seemed to be 'just' clouds of hot gas. But the majority showed patterns that could only have been produced by stars.

But, were these nebulae just collections of stars floating around within the Milky Way, or were they more distant? The answer wouldn't be forthcoming for several decades, so, for the time being at least, the extent of the Universe remained bound up within the Milky Way.

In the 1920s, an American astronomer, Edwin Hubble, finally solved the mystery of Messier's fuzzy objects. (See 'The Cepheid Yardstick', page 23). He proved that they were actually other galaxies located outside of our own galaxy, the Milky Way. The Universe was suddenly a whole lot bigger than we humans had realized.

A DAY WITHOUT YESTERDAY

Although Hubble is often credited with thinking up the idea of an expanding Universe, the true Big Bang daddy was (perhaps a little ironically) a Catholic priest from Belgium called Georges Lemaître.

In 1927, Lemaître proposed that distant galaxies appeared to have been shifted into the red part of the electromagnetic spectrum – ie redshifted – because they are moving away from us, carried away by a Universe that was expanding in all directions.

THE THEORY OF GENERAL RELATIVITY & HUBBLE'S LAW

A man who had never touched a telescope had come to the conclusion that the Universe must be expanding more than a decade earlier than Lemaître. When Albert Einstein formulated his theory of General Relativity in 1916, which describes gravity as the result of mass, energy and the curvature of space-time (we'll get to this later, on page 83), he found the equations were telling him that the Universe had either to be expanding or shrinking, but it couldn't be static. Einstein thought this must be a mistake, so to balance things out he added a bit of mathematical jiggery-pokery to his equations that he called the cosmological constant – a move he would later describe as his 'biggest blunder'. (However, we'll discover later in this book that the cosmological constant was not the cosmological cock-up Einstein thought it was.)

In 1929, Edwin Hubble provided observational evidence of Lemaître's theory of an expanding universe by showing that, relative to Earth, the galaxies were indeed

THE QUICK AND THE RED

When astronomers like Edwin Hubble studied the light spectra from distant galaxies, they noticed that they appeared redder than they should have been. Hubble realized that the further away a galaxy was, the redder it appeared to be.

Light is part of the electromagnetic spectrum and, as such, has a wavelength. Light at the red end of the spectrum has a longer wavelength than light at the blue end.

Longer wavelength   **Shorter wavelength**

Somehow the light from distant galaxies was being stretched into the red end of the spectrum (called redshift). The answer must be that the galaxies were actually moving.

If a galaxy is moving towards the observer, the wavelength of the light it emits is compressed and the galaxy appears blue

If the galaxy is moving away, the wavelength of its light is stretched and the galaxy appears red

The faster the galaxy moves away, the more its light is stretched and the longer the wavelength becomes

The fact that more distant galaxies appeared redder meant that they must be moving away faster than nearby galaxies.

receding. He also showed that the more distant the galaxy, the greater its redshift and the faster it appeared to be moving. From this, Hubble formulated his Redshift Distance Law of Galaxies or, as we know it today, Hubble's Law.

But *how* were the galaxies moving away from us? It is tempting to think that galaxies are whizzing away through space like shrapnel from a bomb, but this is not the case. Hubble's Law, when combined with Einstein's equations, showed that, rather than shooting through space, galaxies were actually being carried by the expanding fabric of space itself (like chocolate chips are carried apart on the surface of a rising cupcake).

AN EXPANDING UNIVERSE

The reason the most distant galaxies showed the most redshift was that their speed increased with distance – the further away they were, the faster they were moving.

This was direct evidence that the Universe must be expanding from a single point (the Big Bang).

1 Imagine the Universe is an expanding bubble.

2 As it expands and carries galaxies with it, those furthest from the observer move away faster and exhibit the most redshift.

3 If the Universe was expanding from a steady state – as if they were being carried along a sort of Universe travelator – all galaxies would move away at the same speed and would display the same amount of redshift.

Several galaxies are visible in this deep image of the Universe. We can tell the galaxy highlighted is further away (and not just small) because it appears much redder than the surrounding galaxies.

THE PRIMEVAL ATOM

Georges Lemaître was the first to propose that galaxies were receding because they were being carried away by a Universe that was expanding in all directions.

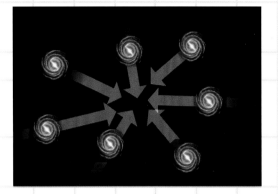

He figured that, if they were moving apart, they must once have been much closer together. So he imagined time running in reverse – with galaxies getting closer and closer, until they converged into a single tiny entity that he called the primeval atom.

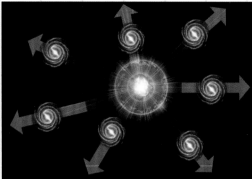

He then restarted his hypothetical clock and imagined the Universe exploding from the primeval atom in what would later be dubbed the Big Bang.

WHAT HAPPENED BEFORE THE BIG BANG?

The short answer to this question is that there was nothing before the Universe 'popped' into existence. But the long answer is a lot more complicated. It is perhaps the first of many bafflingly counterintuitive concepts we will come across in the story of the Universe.

To say something appeared out of 'nothing' would suggest that there had been an absence of 'something' in the first place, but everything was created in the Big Bang: there was no 'nothing' because 'something' had never existed.

We think of 'nothing' as being an absence of 'something' in a region of space (like a bell jar with all the air sucked out of it), but 'space' itself was created in the Big Bang, so there was no framework for 'something' not to exist.

It is just as meaningless to ask what happened before the Big Bang because 'time' didn't exist. There can't be a 'time' before the Big Bang because 'time' was created along with matter and space.

THE PRIMEVAL ATOM

In 1931, Georges Lemaître went on to imagine that, if the galaxies were moving apart, they must, at one time, have been closer together. As he rewound his imaginary Universe clock, he imagined the galaxies getting closer and closer, until they converged into a single tiny entity that he dubbed the primeval atom (except this 'atom' was about 30 times the size of the Sun).

From this concept of a primeval atom, Lemaître imagined the beginning of the Universe as a burst of fireworks, with galaxies as the burning embers spreading out in a growing sphere from its centre. For Lemaître, this burst of fireworks represented the beginning of time, taking place on 'a day without yesterday'.

Despite his brush with an expanding Universe, Albert Einstein was quite disparaging of Lemaître's idea, saying to the Belgian that 'your calculations are correct, but your grasp of physics is abominable'. Eventually, though, he came round to praise it as a 'most beautiful and satisfactory explanation of creation'.

Einstein might have come round to the idea of an expanding Universe born from a primeval atom, but not everyone was so easily convinced. Leading the charge were three astrophysicist pals, Fred Hoyle, Thomas Gold and Hermann Bondi. In 1948, they championed their alternative 'steady state' theory. They argued that, as the Universe expanded, new matter (in the form of stars and galaxies) was continually being created to fill up the gaps. In this way, the Universe could be as uniform today as it was billions of years in the past and as it would be billions of years in the future – it didn't have a beginning, or an end, it just 'was'.

In 1949, while arguing against the primeval atom theory during a radio show, Hoyle disparagingly referred to it as being 'that Big Bang idea'. The name stuck and, from then on, it was known as Big Bang theory.

COSMIC MICROWAVE BACKGROUND

Over the following decades, arguments flowed from one camp to the other, but Big Bang theory steadily gained followers – including the then Pope, Pius XII, who (rather optimistically) believed that it affirmed the idea of a divine creator.

The final nail in steady state's coffin was hammered home in 1964 with the discovery of the cosmic microwave background (CMB), which had been predicted by a Ukrainian-born astrophysicist, George Gamow, in 1948. He had suggested that

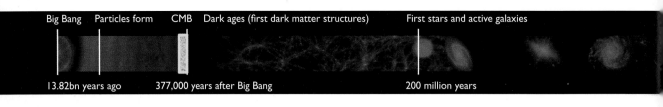

Big Bang   Particles form   CMB   Dark ages (first dark matter structures)   First stars and active galaxies

13.82bn years ago   377,000 years after Big Bang   200 million years

the Big Bang would have created an energy echo that would still exist in the form of background radiation.

When this ancient relic of the Universe's beginning was finally detected (albeit accidentally) in 1964, it confirmed Big Bang theory as the best explanation for the origin of the Universe. In the following decades, it would hold up against every attempt to discredit it and is now considered to be one of the most successful theories in modern science.

However, none of these discoveries would have been possible if astronomers hadn't figured out a way to gauge the scale of the Universe. Without a method for measuring the distance to astronomical objects, we wouldn't know how fast they were receding and wouldn't be able to metaphorically rewind the cosmological clock and bring them all back together to that point, 13.82 billion years ago, when the Universe began. So it's worth taking some time to explore the rather extraordinary series of deductions that showed us ...

HOW TO MEASURE A UNIVERSE

Even as recently as the 19th century, astronomers were hard pushed to tell you the distance to even relatively local objects in space, such as Mars or Venus; the distance to faraway stars or nebulae was anybody's guess.

As we've seen, the invention of the telescope in the 17th century opened up a new frontier of heavenly observation – pinpoints of light that were barely discernible with the naked eye were suddenly revealed as planets, moons, and comets. Even as the Universe seemed to expand before our very eyes, the problem remained that scientists couldn't pace out, drag out a tape measure, or use one of those 'wheel on a stick' things used by neon-clad road workers to figure out the distance between objects. So how exactly did they measure distance in space?

ANGLES OF MERCY

For relatively close matter, the answer can be found in a relatively simple mathematical trick called the trigonometric parallax – to avoid the panic that might be induced by school geometry flashbacks, we'll just call it 'the parallax measurement' (which has a nice 1970s sci-fi thriller ring to it don't you think?).

You can see the parallax effect right now simply by doing the following:

1. Hold a finger a few inches from your nose and close one eye.
2. Take note of the position of your finger relative to a background object.
3. Now close that eye and open the other eye (this trick won't work if you are a Cyclops) and you will see that the finger seems to jump to a different position.

This 'jump' happens because each eye – separated by a distance of a couple of centimetres – sees the finger from a slightly different direction (finger-based parallax experimentation is not recommended in public areas as raised fingers and exaggerated winking behaviour can be misinterpreted).

By measuring this parallax movement and using simple geometry it is possible to

PARALLAX BY FINGER

An easy demonstration of the parallax effect is to hold a finger in front of your face (preferably your own finger).

By opening and closing alternate eyes, you will see the finger appear to jump from side to side. This movement is caused because each eye sees the finger from a different angle.

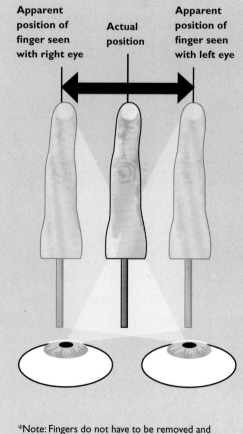

Apparent position of finger seen with right eye   Actual position   Apparent position of finger seen with left eye

*Note: Fingers do not have to be removed and mounted on sticks to achieve this effect.

MEASURING THE PLANETS

By measuring the parallax movement of distant objects against the positions of background stars, astronomers can work out the parallax angle and, by combining this with baseline distance, can use simple trigonometry to calculate the distance to the object.

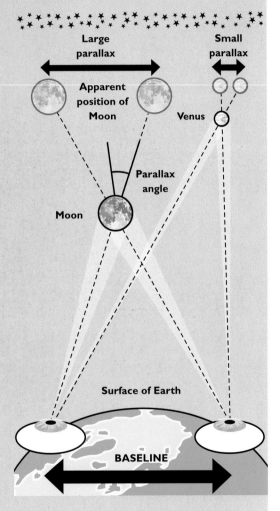

The Moon is close enough for two observatories on Earth to see a significant parallax movement, but the more distant the object is, the smaller the parallax.

work out the distance of the finger from, say, your nose.

The same technique can be used to measure the distance to faraway objects like mountain ranges, the Moon, the planets and even galaxies. Unfortunately (and perhaps unsurprisingly) the further away an object is, the smaller the parallax movement is and the harder it becomes to measure the distance.

Pull out that finger again and pop it back in front of your nose. Now, open and close your eyes again, but this time slowly move your finger away from your face. As you pull it away, you will see the parallax become smaller and smaller. This is because (unless you happen to be a hammerhead shark) your eyes are quite close together and, as your finger moves away, the difference in the angle they are seeing the finger at becomes less and less.

The same is true when astronomers use parallaxes to measure objects in space. To measure the distance to the Moon (only 400,000km away), astronomers need to place their 'eyes' (i.e. two telescopes) a few thousand miles apart. But to measure the distance to even our closest planetary neighbours, Mars and Venus, things become rather more tricky.

Even telescopes placed on opposite sides of the Earth (about 12,000 kilometres) are really close together when compared to the huge distance to Mars (56 million kilometres at its closest), which makes one very skinny triangle. But, although the angles are tiny, they *are* measurable.

THE ASTRONOMICAL UNIT

In 1671, two teams of French astronomers made simultaneous measurements of the position of Mars. One team, led by Giovanni Domenico Cassini, was in Paris, and the other, led by Cassini's assistant, Jean Richer, was sent to French Guiana.

When they met up again, they compared notes and were able to measure the parallax of Mars and, from this, they calculated the distance to Mars. Using this data they were also able to estimate the distance from the Earth to the Sun as 140 million kilometres (the modern measurement is 149.6 million kilometres). This Sun/Earth distance was so important that it has become astronomy's standard unit of measurement for objects in the Solar System: the astronomical unit, or AU.

But why was the AU so important? Well, it gave astronomers a new baseline from which to measure the Universe. Since the radius of Earth's orbit is almost 150 million kilometres, observations made six months apart, from opposite sides of the Sun, give astronomers a baseline of 300 million kilometres – with 'eyes' that far apart it is possible to measure the distance to objects far outside of our Solar System.

Unfortunately for astronomers, even with a baseline of 300 million kilometres, the parallax of even 'nearby' stars is really very small – too small for the telescopes of the 17th century to resolve. To make matters worse, astronomers hadn't figured out how to compensate for the 'wobble' of Earth's axis (called nutation) and its motion as it travels through its orbit, which causes the light from stars to appear to strike the

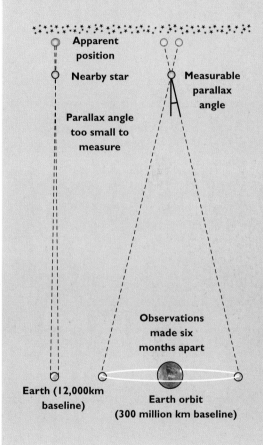

MEASURING THE STARS

By taking measurements six months apart, when the Earth is on opposing sides of its orbit, astronomers can increase the length of their baseline – allowing the parallax to more distant objects to be measured.

Apparent position

Nearby star

Measurable parallax angle

Parallax angle too small to measure

Observations made six months apart

Earth (12,000km baseline)

Earth orbit (300 million km baseline)

Earth at an angle (like the way rain falling straight from the sky seems to 'blow' into your face when you run forward) – an effect called 'aberration'.

It would take more than 150 years for theory and telescope technology to reach the point where astronomers could measure the distance to the stars. But, following a flurry of astronomical activity in the 1830s, by the mid-19th century astronomers had charted the distance to several of the Sun's closest neighbours (closeness in astronomy is all relative of course – even the nearest star, Proxima Centauri, is located 271,000AU away ... that's 271,000 × 150 million kilometres, or 39.9 trillion kilometres).

But once again astronomers ran up against the brick wall of technological limitations and, after the initial parallax gold rush, measurements ground to a halt. It was clear that astronomers had reached the limits of what they could measure by parallax alone – it's a bit like knowing the distance between two houses in a cul-de-sac: it allows you to measure the rest of the street and even estimate the dimensions of the village but it is not much help when trying to work out the distance to the next village, and is next to useless for figuring out the scale of the country as a whole. So clearly, to figure out the size of the Universe as a whole, they needed to come up with something else.

It was only in the early 20th century, with the development of photographic plates, that the field could continue to develop significantly. Before the advent of photography, even the best telescopes were limited by a crucial handicap – the human eye. The teeny tiny parallax of faraway stars was just too small for the human eye to register.

Photography's most obvious advantage was that it provided a permanent, accurate record of star positions, which could be studied at the astronomer's leisure when he, or she, wasn't shivering away in a mountain-top telescope. The positions of stars could be measured with great precision – even under the gaze of a microscope if needed.

Its biggest advantage was that the longer the plate was left exposed, the more light fell on it, and even faint images became brighter. With the human eye, you can stare into the heavens for as long as you want, but you won't be able to see a faint object any more clearly than when you started.

Before the advent of photographic astronomy in 1900, the parallaxes of just 60 stars had been measured. It took just 50 years for that number to increase to almost 10,000. The increasing abundance of measured stars allowed astronomers to create a catalogue of star attributes that they could use to estimate the distance to stars that were too remote to be measured by parallax directly.

STAR LIGHT, STAR BRIGHT

At the start of the 20th century, astronomers had identified a link between a star's colour and temperature, and its brightness. Using the spectroscopy techniques pioneered by William Huggins in 1860, a Danish astronomer, Ejnar Hertzsprung, and an American, Henry Norris Russell, were able to determine quite independently of each other that the vast majority (about 90 per cent) of stars fall into a neat colour range that spreads from blue to red.

Blue stars are just starting out in life and burn with such ferocity that they appear blue (like the hottest part of a flame is blue) and red stars are stellar-geriatrics that burn at a cooler, more leisurely pace (our Sun is right in the middle of that range and so appears yellow).

The brightness, or luminosity, of a star is also directly connected to its temperature – hotter stars produce more light and so shine brighter. By combining the spectroscopy readings with distances obtained using parallaxes, Hertzsprung and Russell were able to create a graph that showed how bright each kind of star should be. This is called (perhaps unsurprisingly) the Hertzsprung–Russell diagram.

Light (and all electromagnetic radiation) obeys something called the inverse-square rule, which basically means that for every unit of distance, the brightness of a star decreases by the square of that distance – so, a star two units away appears four times dimmer (2×2) and a star four units away appears sixteen times dimmer (4×4).

Now, when a star was discovered that was too far away to use the parallax method, all an astronomer had to do was identify what kind of star it was using spectroscopy, match it to the Hertzsprung–Russell diagram, compare its apparent brightness and use the inverse-square rule to estimate its distance (it's like knowing the luminosity of a 60 watt light bulb and using that knowledge to estimate how far away another 60 watt bulb is).

This method of determining distance is called the spectroscopic parallax technique (which is slightly confusing as it has nothing to do with parallax), but even this can only help us measure distance to relatively nearby stars.

THE CEPHEID YARDSTICK

The further starlight has to travel to reach us, the more light-obscuring 'stuff' can get in its way, such as dust, which absorbs and reflects light. The light that does eventually

MEASURING THE GALAXY

Once a link was found between a star's colour (spectral barcode) and its brightness, astronomers just had to look for stars with the same properties as a star they already knew the distance to. Since they knew how bright it should be, they could use the inverse-square rule to estimate is distance.

Nearby star (distance known)

Distant star (with matching spectral barcode)

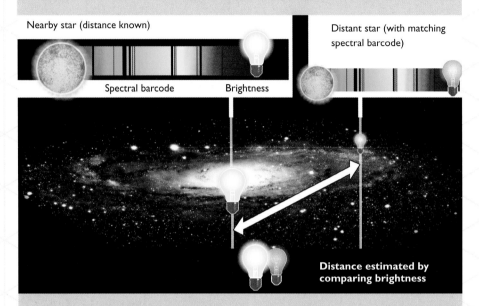

Spectral barcode

Brightness

Distance estimated by comparing brightness

THE INVERSE SQUARE RULE

As light travels through space, it spreads out in a sphere. Since the number of photons remains the same, the further the light travels, the fewer photons will occupy any given area. Photons twice as far from the light source are spread across four times the area, hence appearing one-fourth the brightness.

Distance: 1 unit

2 units

3 units

1/4

1/9

Light spread over 4 times the area

Light spread over 9 times the area

make it through these obstructions can't be trusted to be telling the 'truth' about the star it came from. The problem is that the light gets absorbed and remitted by atoms of dust, which alters the spectrum of the light by the time it reaches Earth – the spectra of 'all the stuff in-between' gets mixed up with the pure spectra of the light as it was originally emitted.

To make the final step to measure outside our galaxy, astronomers needed to find the cosmic equivalent of a road sign – a single object that could cut through all the 'stuff' or 'noise' and be used as an intergalactic mile marker. They didn't have to wait long for one to be discovered.

First spotted in 1784, Cepheid variable stars spent more than two

MEASURING THE UNIVERSE

It takes a very special sort of star to cut through the 'noise' created by dust and gas in intergalactic space. A Cepheid variable is just such an object.

Cepheid variable stars swell and contract and, in doing so, vary in brightness (luminosity) – pulsing from bright to dim and back to bright again over a measurable period.

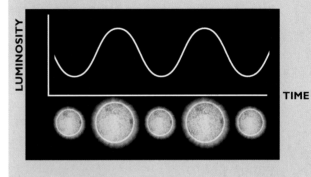

The brightness of a Cepheid is connected with its period. By studying a Cepheid's period, astronomers can determine its brightness and, by applying the inverse-square law, estimate the distance to the galaxy in which it lives.

centuries filed under the 'interesting oddities' section of astronomers' minds. As the name suggests, variable stars are stars that vary in brightness – throbbing from bright to dim and back to bright again like twinkling Christmas tree lights. And like those festive decorations, Cepheid variables can pulse away at all sorts of different rates – some pulse slowly, some fast, and some pulse fast, then slow, then fast again.

But it wasn't until the first decade of the 20th century that the potential for Cepheids to act as cosmic yardsticks was discovered by a 'computer' at Harvard University.

These were the days before computers were rooms full of glowing valves and spinning data tapes – instead, computers were women employed to catalogue the brightness of stars recorded on photographic plates (in those days, women weren't trusted to operate the complex and expensive telescopes).

One such computer was Henrietta Swan Leavitt. In 1908, she discovered that there was a predictable link between the brightness of Cepheid stars and the period of their variation (in other words, a Cepheid that varies from bright to dim to bright over a period of two days will have a different brightness to one that has a period of seven days). If astronomers could find the distance to a Cepheid of a known period (which they did, in 1912, using the techniques we have already discussed), then all they had to do was find another Cepheid with the same period, measure its brightness, apply the inverse-square law, and voila!

Suddenly, astronomers had a standard yardstick they could use to measure almost any distance in the Universe (later, supernova explosions would be used to measure the most extreme distances). For this reason, Cepheids are known as standard candles.

Two decades later, Edwin Hubble used Cepheid measurements to prove that some of Messier's 'fuzzy blobs' were too far away to be within the Milky Way and had to be separate galaxies. He measured the distance to our nearest galactic neighbour, the Andromeda Galaxy, and determined that it was some 800,000 light years away (one light year is about 10 trillion kilometres). With that single conclusion, the size of the Universe as we understood it expanded well beyond the confines of our galaxy. Cepheids helped reveal that, far from being unique, our galaxy is just one of countless billions in a Universe that, from our standpoint, had just increased exponentially in size.

LIGHT YEAR

Until the mid-1800s, the largest unit of measure was the Astronomical Unit (about 150 million km – the distance between the Sun and Earth). When astronomers set out measure the distances to the stars, they realised that they needed a more suitable measurement. The speed of light (about 671 million mph) had been calculated in 1729 by English Astronomer, James Bradley. In 1838, a German astronomer, Friedrich Bessel, used this figure to calculate how far light would travel in a year and used the 'light year' to describe the distance to a star called 61 Cyni (also known as 'Bessel's star').

SIZE IS EVERYTHING

Hubble's 800,000 light-year distance to Andromeda might sound a lot, but when he used the distance as a yard stick for other galaxies, there was a problem. When he applied his Redshift Law to figure the speed that galaxies were moving away and wind the clock back to the Universe's start-point, the figures implied that the Universe was only two billion years old, which was just too young.

This caused a colossal headache for supporters of the Big Bang model. Geologists had already determined, by studying rocks and meteorites, that the Earth and Moon were at least four billion years old. Obviously the planets couldn't be twice the age of the Universe – something was wrong. The Universe had to be larger than it appeared to be (if it wasn't, Big Bang theory would have died there and then).

The early attempts of the likes of Hubble to estimate the distance to the nearest galaxies were hampered by the inability of the telescopes of the day to resolve them as anything more than fuzzy little blobs. It wasn't that the telescopes were not powerful enough, but they were critically handicapped by the distorting effects of the Earth's atmosphere and by man-made light pollution.

Light pollution, particularly in cities, is a huge problem for astronomers today. Light photons, which have travelled billions of miles across space from a star that, from Earth, appears no bigger than a pin head seen from two miles away, are easily drowned out by the millions of artificial lights that spew photons into the skies above a city. Luckily (for our story at least), in the 1940s, astronomers found themselves with an unlikely ally: the Second World War.

The fear of night-time bombing meant that black-outs were common; after all, why provide enemy bombers with a brightly lit target to aim for? With the flick of a switch, the blinding orange glow of artificial light was removed from skies and one astronomer was ideally placed to take advantage of it.

The German-born astronomer, Walter Baade, had moved to the United States in 1931 to escape his homeland's increasingly unstable political climate. When the US became involved in the conflict, most of his scientific colleagues were diverted into military research, but as a German national and potential security risk Baade was left out of the war effort.

He found himself with a virtual monopoly of what was then the biggest telescope in the world: the 100-inch Hooker Telescope on Mount Wilson, California (used by Hubble to discover his redshift law). Best of all, Baade had the luxury of using it to explore truly dark skies.

Exploiting every inch of the telescope's power, Baade was able to resolve the Andromeda Galaxy (our nearest large galactic neighbour), that Hubble had only been able to see as a fuzzy blob, into individual stars.

AGE OF THE UNIVERSE

The age of the Universe was most recently refined by the European Space Agency's Planck spacecraft. The craft has mapped the radiation 'afterglow' of the Big Bang (the Cosmic Microwave Background, or CMB) with an unprecedented level of accuracy that, in 2013, allowed cosmologists to more accurately define the Hubble constant (Hubble's Law) – pushing back the birth of the cosmos from 13.73 billion years (estimated using data from Nasa's CMB probe, WMAP) to 13.82 billion years.

He discovered that it contained two distinct types of Cepheid star. The first, called Population I, are young stars that are hot and blue, and the other, called Population II, are old stars that are larger, cooler and redder.

When Hubble had originally calculated the distance to the Andromeda galaxy using the brightness of the fuzzy blob, he had been unaware of the dimmer Population II stars. So he calibrated his calculations just using the much brighter Population I stars – and because they were brighter, they made it appear as if Andromeda was much closer than it really was (in essence, he thought he was looking at a 60 watt bulb when he was really looking at a 100 watt bulb).

When Baade recalculated the distance to Andromeda using the correct luminosity, he came up with a figure of 2 million light years (modern estimates up it to 2.5 million light years). With a click of his mathematic fingers, Baade had doubled the distance to every galaxy outside of the Milky Way and increased the age of the Universe to about five billion years, which, although still a long way short of today's estimate of 13.82 billion years, at least meant the Universe was older than our Solar System (which was a great relief to Big Bang astronomers).

OK, so that's how we came to discover that the Universe is a helluva lot bigger than anyone had imagined and how it all kicked off with something called the Big Bang. Now it's time to move on and start thinking about how we might set about building a Universe (leave delusions of divine omnipotence at the door please). We know that it started with a Big Bang, so it's logical that we start with a Big Bang, which, as it turns out, isn't big and involves no kind of bang whatsoever.

THE MOUNT WILSON
TELESCOPE

Above: Then the largest telescope in the world, the 100-inch Hooker
Telescope at the Mount Wilson Observatory, Los Angeles, California,
was used by Edwin Hubble to formulate his redshift law to measure the
expansion of the Universe in 1929. But he was hampered by the fact he
couldn't resolve individual stars within the galaxy and had to take his
measurements from a 'fuzzy blob' (inset picture).

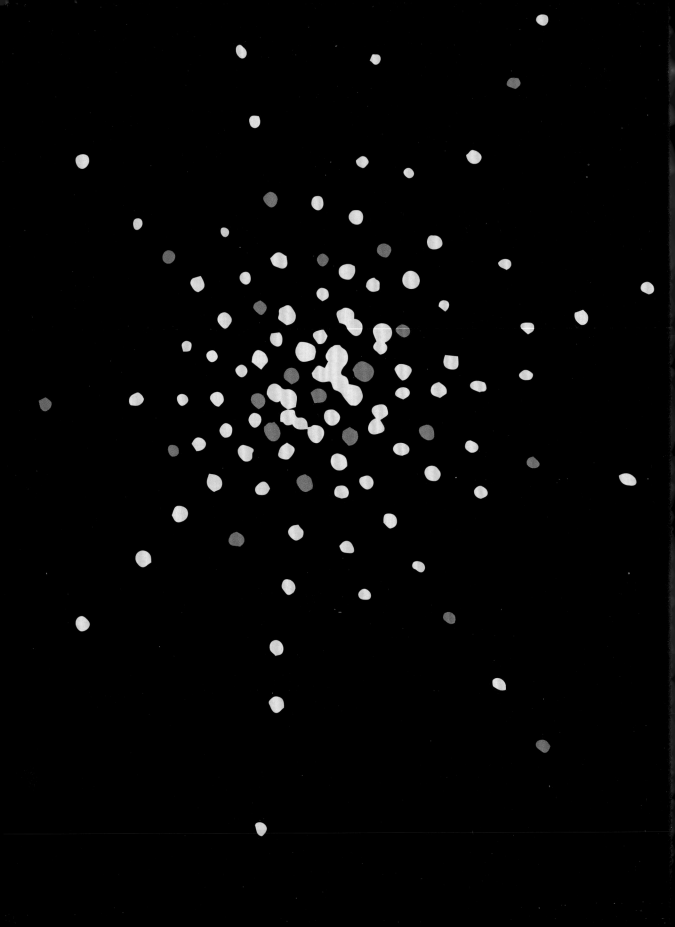

A UNIVERSE IS BORN

In which we chart the birth of the Universe and the formation of the first particles of matter that we will use to build our Universe.

In the beginning God created the Heavens and Earth ...'
So begins the Christian view of the start of all things: nine
words to explain how the Earth and the Universe came
into existence. It seems a bit simplistic doesn't it? Hardly a
thorough analysis of such a monumental occurrence.

But how does modern science explain how the Universe came into existence? It
must be a very long and complicated account with lots of complicated jargon such as
'entropy', 'homogeneity' and 'isotropy', no? Well, science describes the moment of
'creation' even more laconically than the Bible does – read a science book (a bit like
this one) and they will all say that the Universe was born in ... 'the Big Bang'. Just
three little words: three trivial, monosyllabic and child-like words to describe how a
Universe of almost incalculable size was born.

A DAY WITHOUT YESTERDAY

IS THERE A CENTRE OF THE UNIVERSE?

There is no centre of the Universe.
It isn't expanding from a fixed central
point but instead is expanding equally
from every point. No matter where
you are in the Universe it would appear
that all of space was expanding away
from you and that you are at the centre
of the cosmos.

OK, so the initial science account is a
little disappointing but, compared to
the Biblical story of a supernatural deity
summoning a Universe from his heavenly
pockets, surely the scientific account of
how the Universe came to exist will deal
in concrete evidence – actual things we
can see, hear and conduct repeatable
experiments on?

Er, no ... Not really.

The trouble with Big Bang theory
is that it doesn't really describe how the
Universe was born. Perhaps misleadingly,
the term Big Bang doesn't refer to a
cosmic explosion from which space, time,
and matter was born. Instead, it describes

the evolution of the Universe *after* it came into existence.

Think of Big Bang theory as being an account of your life written by an alien with no knowledge of human physiology. The alien will be able to accurately describe how you develop and interact with your environment from the moment of your birth, but it can only guess what happened to cause your creation in the first place.

Big Bang theory can describe the development of the Universe from just a fraction (of a fraction, of a fraction, of a fraction) of a second after its birth – but it can't explain how it came to exist in the first place.

So, as it stands, science can't explain how the Universe came into existence – did it just 'pop' out of nothing (as if pulled from God's pocket)? Is it part of a continuous cycle of life, death and rebirth? Is the Universe even unique? Could it be just one bubble in a colossal 'Swiss cheese' Megaverse? We will get to these questions later but, for now, we are getting ahead of ourselves.

WHAT CAN BIG BANG THEORY TELL US ABOUT THE UNIVERSE?

So, physics can tell us a lot about the birth of the Universe but it can't explain what happened in the first infinitesimally small fraction of a second. Here, in what is known as the Planck era, the laws of physics break down and theoreticians find themselves wrestling with equations that spit out numbers, which spiral into infinities (they hate infinities) – there is nothing in classical or quantum physics that can explain what's going on. If Big Bang theory were a book, it would start at the second paragraph on the first page of the first chapter and, although we have to wait for scientists to write that first paragraph, we can still enjoy the rest of the book.

Our story starts with something smaller than the smallest particle. Crammed within this smallest of small things is all the matter and energy that will ever exist in the entire history of the Universe – all the galaxies, stars, planets, moons, and life that have ever been, or ever will be; squashed into a dot of potential, just 100,000,000,000,000,000,000 times smaller than a single proton. Then it starts to expand.

THE ATOM

We'll look at what makes up an atom in much more detail later, but for now here's a basic model of the atom.

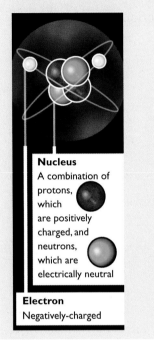

Nucleus
A combination of protons, which are positively charged, and neutrons, which are electrically neutral

Electron
Negatively-charged

In the first few moments of its life (0.0000 0000000000000000000000000000000000 0001 seconds after it was born), the Universe was extremely dense and absurdly hot. Of course, to say it was hot is a bit of an understatement – the centre of the Sun, which boils away at about 10 million degrees Celsius, is 'hot', but you need to add another 19 zeros to that figure to get anywhere near the temperature of the early Universe.

All of this energy (heat and energy can be considered to be the same thing) meant that matter as we know it couldn't form and all the fundamental forces that would one day lend structure to the Universe – electromagnetism, the strong and weak nuclear forces, and gravity (see Chapter 4 – The Force is Strong with this Universe, page 70) – were bundled together as a single unified force.

For the first fraction of a second of its life, the Universe expanded relatively slowly, but, as it did so, it became ever so slightly less dense and ever so slightly less hot. Then something extremely dramatic happened: the unified force suddenly broke down and the fundamental forces split away from each other, and, in doing so, released a colossal burst of energy (a bit like opening a can of shaken-up cola). This energy injection caused the Universe to expand exponentially in a process known as cosmic inflation. In just 0.000000000000000000000000000000000 01 seconds, it expanded by a factor of 10^{78} (that's ten multiplied by ten 78 times over) and that smaller-than-a-proton speck expanded to the size of a grapefruit (equivalent to expanding a tennis ball to the size of today's observable universe) and, by the time inflation had petered out a moment later, the Universe was only about 1,000 times smaller than it is today.

In this far roomier Universe, all the energy that had been so tightly squashed together was able to spread out and, because there was less energy in any given area, it cooled down – allowing the first particles of matter to form. (See page 69 to find out how matter can form from energy.)

COSMIC INFLATION

Cosmic inflation might sound like it's a bit of a Band Aid solution to patch up the shortcomings of the Big Bang model, but a recent discovery has pushed inflation from the darkness of the theoretical fringes towards the spotlight of near-certitude.

One of the predicted side-effects of cosmic inflation is that the brief, but violent, expansion of the Universe would create waves in the fabric of space-time that would have propagated out through the expanding cosmos, like ripples across a pond after a stone has been hurled in.

Known as gravitational waves, it was predicted that these ripples would leave a tell-tale signature in the radiation soup of the infant cosmos. Because, as the waves travelled though space, bending and stretching the fabric of the Universe, they would have bent and stretched any particles that sat within it.

It is in the nature of some particles, such as electrons, to shed unwanted energy in the form of photons (packets of electromagnetic energy, such as light). Under normal circumstances, these photons would be emitted in random directions, but, in the presence of the distorting effects of gravitational waves, they become aligned with the motion of the wave and the photons become polarized.

Lots of things can polarize photons (such as light bouncing from a wet road), but those polarized by gravitational waves were predicted to form a distinctive curled pattern, known as B-mode polarization. If this 'curl' was found to be present in the radiation afterglow of the Big Bang (the Cosmic Microwave Background, CMB) it would be evidence that cosmic inflation did indeed happen – because no other phenomenon capable of distorting spacetime in that way existed at this time.

In 2014, a team working on a telescope based at the South Pole, called BICEP2 (Background Imaging of Cosmic Extragalactic Polarization experiment) announced to a very excited cosmologist community the discovery of this very particular polarization fingerprint hidden within the CMB.

The result is far from being 'iron clad' and, as with all scientific discoveries, it will need to be independently verified by other experiments, but, for now, cosmic inflation is the best tool we have to help us build our Universe.

MATTER VERSUS ANTIMATTER

The very first particles to congeal from the energy soup were the fundamental particles (sometimes referred to as the elementary particles). These included the photons (packets of light energy); the electrons (packets of electromagnetic energy); and the quarks (tiny chunks of matter).

Created at the same time was an almost equal quantity of antimatter. Antimatter particles exactly match the characteristics of their matter brothers, except their electrical charge is reversed (the antimatter opposite of the negatively-charged electron is the positively-charged positron).

Another peculiarity of antimatter is that it really doesn't get along with matter and if they try to occupy the same space they annihilate each other in a reaction that instantly converts all of their mass back into energy. Because the early Universe was so dense, matter and antimatter didn't have the luxury of being able to avoid each other, so vast quantities of all that freshly made matter was converted back into energy during matter/antimatter reactions.

All this re-liberated energy served to further drive the expansion of the Universe but, eventually, things calmed down enough for the energy to reform back into the fundamental particles of matter. The construction of the Universe could continue.

We can be extremely grateful that matter and antimatter weren't created in equal measures. If the quantities had been perfectly balanced, they would have continued to wipe each other out until neither one remained and the Universe as we know it would have been snuffed out there and then.

MATTER/ANTIMATTER ANNIHILATION

When an electron collides with its antimatter opposite, the positron, all of their mass is converted into energy.

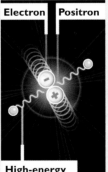

Electron | Positron

High-energy photon

Matter/antimatter reactions release so much energy that they could be harnessed to power future generations of space explorers to the planets in record speed.

Just 10 milligrams of positrons can unleash the energy of 428 tonnes of TNT. That's as much energy as 23 external fuel tanks (the big orange ones) on the Space Shuttle.

THE QUARK

All the fundamental particles are referred to as the building blocks of matter, but (although photons and electrons are very important) the true building blocks are quarks because (in different combinations) they make up the protons and neutrons that form the atomic nucleus.

Quarks are peculiar little chaps to say the least. While most particles are happy to carry a nice even electrical charge (the electron is -1, proton +1, etc.), quarks only carry a fraction of a charge. There are three quarks that carry a ⅔ positive charge and three 'partner' quarks that carry a ⅓ negative charge. They are also suitably eccentrically named. There's the up quark and its partner the down quark; there are the charm and the strange quarks; and then there are the top and the bottom quarks.

Quarks are also unusually social particles. Never found alone, they are always bonded in pairs or triplets* from which they refuse to be separated. In fact, science has never seen (nor will it ever see) a lone quark. They are held together by the strong nuclear force (see page 72), which on atomic scales actually gets stronger over distance. If you were to get out a tiny atomic crowbar and try to prise two quarks apart, you would feel the force attracting them to each other get stronger and stronger (rather like the resistance of a rubber band increasing when you stretch it). Eventually, so much energy would have been put into pulling them apart that the energy is converted into mass and you will find yourself with two brand new (and equally firmly bonded) quarks.

There is no process that we know of today that can provide the energy needed to overcome the quark's attraction (well, a particle accelerator the size of the Sun might do it) but, in the insanely hot particle soup of the infant Universe, quarks roamed free, which shows just how much energy was swimming around in those first moments.

But their freedom was short-lived and, just 0.01 seconds after the Universe's birth, the soup had cooled enough (to a pleasant 10 million million degrees) to allow the quarks to be captured by their mutual attraction and become bonded together to create the first neutrons and protons – the quark would never be free again.

*... possibly even quadruplets! In 2014, scientists at the Large Hadron Collider found strong evidence of a particle made up of four quarks. If confirmed, this 'tetraquark' particle would represent a new form of matter and could mean there are whole families of pentaquarks and hexaquarks hiding out there too – possibly within a hypothetical stellar remnant called a quark star.

THE MAKING OF MATTER

Time	**1** Planck era 13.8 billion years ago	**2** Fundamental particles 0.00000000000000000000000000000000 00000000000000001 seconds later	**3** Protons and neutrons 0.0000001 seconds later

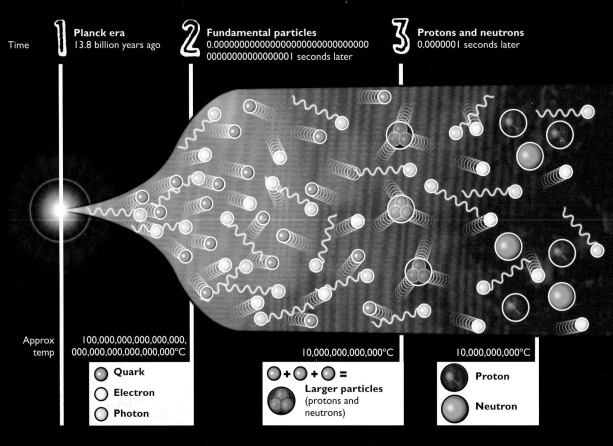

Approx temp	100,000,000,000,000,000, 000,000,000,000,000,000°C	10,000,000,000,000°C	10,000,000,000°C

Quark

Electron

Photon

○ + ○ + ○ =
Larger particles
(protons and
neutrons)

Proton

Neutron

1 Planck era: Space, time, matter, and energy are all bundled up in an impossibly small, infinitely dense, insanely hot fireball. All the fundamental forces (gravity, electromagnetism and the strong and weak nuclear forces) are also bundled together as a single unified force. A trillionth of a second later, the unified force breaks down and powers the exponential inflation of the Universe.

2 Fundamental particles: As it expands, all that energy becomes less dense and the Universe cools down. Energy congeals into matter and the first particles are born. These first particle building blocks – quarks, electrons, photons, and neutrinos – are created along with their antimatter twins (antiquarks, positrons, etc). These matter opposites collide and annihilate each other, releasing huge numbers of photons (light particles).

3 Protons and neutrons: As the temperature drops, colliding quarks can join together without being torn apart immediately by all that energy. Quarks combine (via the strong nuclear force) in sets of three to form the first protons and neutrons.

Big Bang   Particles form   CMB   Dark ages (first dark matter structures)   First stars and active galaxies

13.82bn years ago   377,000 years after Big Bang   200 million years

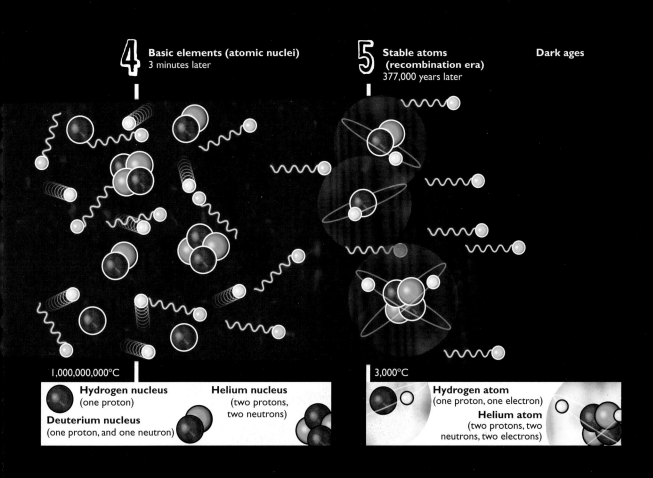

4 Basic elements (atomic nuclei)
3 minutes later

5 Stable atoms (recombination era)
377,000 years later

Dark ages

1,000,000,000°C

Hydrogen nucleus (one proton)

Helium nucleus (two protons, two neutrons)

Deuterium nucleus (one proton, and one neutron)

3,000°C

Hydrogen atom (one proton, one electron)

Helium atom (two protons, two neutrons, two electrons)

4 Basic elements: When the temperature has dropped to about a billion degrees, colliding protons and neutrons can combine through nuclear fusion to form the nuclei of the simplest chemical elements – hydrogen, deuterium (heavy hydrogen) and helium. About 20 minutes later, the Universe has cooled too much and nuclear fusion ends (it won't start again until the first stars are born). During this era, the Universe is filled with a hot, opaque soup of atomic nuclei and electrons called plasma. All of the photons created through matter/antimatter annihilations are trapped within the plasma – forever colliding with protons and electrons.

5 Stable atoms: The Universe cools enough to allow the positively-charged atomic nuclei to capture the negatively-charged electrons – becoming neutral. With all the nuclei stabilized, photons can travel unimpeded and the Universe becomes transparent for the first time. At this point, the Universe is made up of 75 per cent hydrogen and 25 per cent helium.

Galaxies evolve (clusters and superclusters form)

Solar system forms

Death of the Sun

Fate of the Universe

1 billion years

9 billion years

18.7 billion years

THE RADIATION ERA

It's worth taking a moment to try to visualize what the Universe was like at this point. Even though it has cooled down a great deal, it is still about a thousand million degrees (a billion in new money), which is still *really quite hot*. Heat is just another way of describing how much energy there is within a system – when particles are hot, they have a lot of energy so they fly around very quickly (the colder something is, the less energy it has and the slower its atoms move around – which is why hot water becomes steam and why cold water becomes ice).

At a thousand million degrees, there are a lot of particles, with a lot of energy, moving around very quickly. Because the Universe is still a relatively dense soup of particles, all those energetic particles are constantly smashing into each other and, every so often, a

IMPRISONING LIGHT

According to Einstein's theory of Special Relativity, the speed of light is constant at 299,792,458 metres per second (about 670 million miles per hour). For example, the light that leaves the surface of the Sun takes just 8 minutes to cover the 93 million miles (150 million km) to Earth. But the crucial part of that statement is 'leaves the surface of the Sun' – the speed of light might be constant in the vacuum of space but stick enough obstacles in the way and it can take an awful lot longer to get from A to B.

The photons that reach us might have only taken a few minutes to cover the ground between our local star and Earth but they

actually started their journey quite a lot earlier. The moment they were created by nuclear fusion in the Sun's core, they faced a problem that the Universe's earliest photons would sympathize with. Each photon produced in the core has to claw its way through 430,000 miles of extremely dense plasma. It can only travel a tiny distance before it runs into a hydrogen nucleus and gets absorbed and re-emitted in a random direction. A photon might get absorbed, emitted, re-absorbed and re-emitted trillions of times (a process affectionately known as the 'drunkard's walk') before it finally reaches the Sun's surface up to 170,000 years after it began its journey.

proton will collide with a neutron and they will stick together – creating the first nuclei of light elements.

Less than three minutes after its birth, the Universe now has its first elements – hydrogen (one proton), deuterium (an isotype of hydrogen with one proton and one neutron), helium (two protons and two neutrons) and a very small amount of lithium (three protons and three neutrons). But there is still too much energy in the system for those atomic nuclei to capture and keep hold of the electrons they need to become stable atoms – all those particles flying around just knock them off again. When atoms are lacking their orbiting electrons they are known as ions, or charged particles, because (somewhat obviously) they carry an electrical charge. A fully fledged hydrogen atom, for example, has one negatively-charged electron orbiting one positively-charged proton nucleus – positive cancels out negative and you have one neutral atom. A neutral helium atom has two protons and two neutrons (neutrons, as their name suggests, carry no charge) in its nucleus, which is orbited by two electrons.

So, instead of a nice stable Universe of neutral atoms, what we have at this point is a dense roiling ocean of super-heated plasma – hot electrically-charged gas – not unlike the Sun. In effect, the Universe is like a giant star. All this particle activity means that light can't travel very far. There are so many particles and free electrons whizzing around that light photons are continually blocked, absorbed and reflected.

Unlike all the other phases in our story so far, the radiation era doesn't pass in the blink of an eye but lasts for almost 380,000 years. Everything that happened from the appearance of that first smaller-than-a-proton Universe seed to the end of the radiation era has to be inferred from theory and experimental evidence, and is the exclusive domain of physicists.

All that is about to change: light will be released from its primitive plasma prison and observational astronomers will be able to join the party. But, before this can happen, something very important has to take place: recombination.

DRUNKARD'S WALK

Light travels very fast in a straight line, but in the dense early Universe, a straight line wasn't an option.

Charged particles

Photon absorbed and emitted in random directions

HOW DO ASTRONOMERS 'LOOK' INTO THE PAST?

Because light takes time to reach our eyes (or the lens of a telescope), the more distant an object is, the further back in time we are seeing it. Light from the most distant star visible with the naked eye is about 16,000 light years from Earth, which means we are seeing it as it was during our stone age – it could explode today and we wouldn't know about it for another 16,000 years.

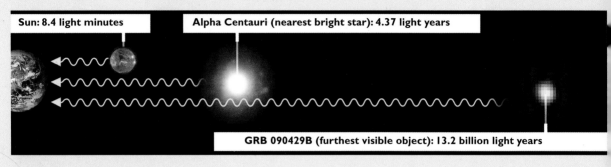

Sun: 8.4 light minutes

Alpha Centauri (nearest bright star): 4.37 light years

GRB 090429B (furthest visible object): 13.2 billion light years

Light travels in a vacuum at 186,000 miles per second (300,000 kilometers per second) – that's 7.5 times around the Earth every second. A light year is the distance light can travel in a year, which means a light year is 5,865,696,000,000 miles (9,460,800,000,000 kilometers).

THE RECOMBINATION ERA

The recombination era marks a point in the Universe's story when it had cooled sufficiently (to about 3,000°C) to allow the charged particles of hydrogen and helium to sweep up all those free electrons and lock them away into nice stable orbits.

Recombination is a slightly misleading term – after all, those atomic nuclei had never been combined with electrons in the first place (it's rather like the March Hare asking Alice if she wants more tea when she hadn't had any to start with). Yet, for reasons known only to astrophysicists, this crucial turning point in the story of the Universe is known as the recombination era.

At last, the stable atom as we know it had been born and for the first time light could travel freely through the Universe. For astronomers, it represents the moment a giant metaphorical lens cap was removed from the cosmos.

Almost 14 billion years after the lens cap was removed, this 'first light' was detected by telescopes on Earth and astronomers called it the cosmic microwave background (CMB). Because the light was released almost instantly from the entire Universe, it provides a perfect snapshot of the infant Universe, allowing us to map its make-up when it was just 380,000 years old (see opposite).

PATCHING UP THE BIG BANG

Earlier in this chapter we talked about a short period of exponential expansion called cosmic inflation. Today, this idea has been accepted by most cosmologists, but when it was thought up in the 1980s, inflation was a rather radical idea designed to plug a troublesome hole that the discovery of the cosmic microwave background had blown in Big Bang theory.

In the 1940s, Big Bang physicists had predicted that, if the Universe had indeed been born as a maelstrom of high-energy radiation, during the following billions of years of expansion, that early radiation would have cooled and become stretched into the microwave region of the electromagnetic spectrum. If we could look back far enough through the electromagnetic 'noise' of the Universe, we should be able to detect those ancient photons as a sort of microwave canvas on which the Universe was painted.

HOW CAN WE 'SEE' BACK IN TIME?

Every time you open your eyes to look at the world around you, you are looking into the past.

Light travels at 300,000 kilometres per second, which is very fast, but because it has to travel from an object to your eyes, you are actually seeing that object as it appeared when the light left it – you are looking into the past.

Light travels so fast that, for nearby objects, the time delay is virtually nothing (about a billionth of a second for an object one metre away), but the further away the object is, the longer light takes to make that journey and the further back in time you can look.

Light from the Sun has to cover a distance of 150 million kilometres before it reaches our eyes here on Earth, so, we are seeing the Sun as it was about 8.4 minutes in the past.

One of the most distant stars visible to the naked eye is V762 Cas in the Cassiopeia constellation. It is about 154 million billion kilometres away, which is 16,308 light years; this means we see this star as it was more than 16,000 years ago (when the light from this star began its journey, we humans were right in the middle of our stone age).

This time delay is taken to extremes when you use a telescope. The most distant object modern telescopes can see is GRB 090429B a gamma ray burst whose light took 13.2 billion years to reach us.

ECHOES OF THE BIG BANG
THE UNIVERSE'S FIRST BABY PICTURE

This is the cosmic microwave background (CMB) radiation as seen by the European Space Agency's Planck spacecraft. It dates from about 380,000 years after the Big Bang and represents the first light of the Universe – released when it had cooled enough for neutral atoms of hydrogen and helium to form and allow photons of light to travel unimpeded through space for the first time.

The CMB map shows deviations in the average temperature of the early Universe. Although the colour differences look dramatic (blue is colder and red is warmer) they actually represent temperature differences of less than a hundred millionth of a degree.

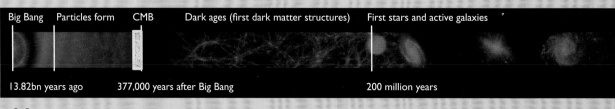

Big Bang	Particles form	CMB	Dark ages (first dark matter structures)	First stars and active galaxies
13.82bn years ago		377,000 years after Big Bang		200 million years

THE ELECTROMAGNETIC SPECTRUM

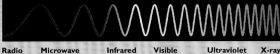

Radio   Microwave   Infrared   Visible   Ultraviolet   X-ray

When the light from the CMB began its journey, the Universe was still about the same temperature as the surface of the Sun today and the CMB was emitted as heat – also known as infrared radiation. But, as the Universe expanded, the wavelength of the light was stretched into longer, cooler wavelengths (a bit like a wavy line drawn on an elastic band becomes stretched when the band is pulled).

The CMB reveals how evenly spread matter and energy were in the early universe – it barely fluctuates from 2.7 degrees above absolute zero (-273C).

This uniformity of temperature couldn't have been created by a Universe that expanded slowly, so is seen as evidence that Universe underwent the period of exponential expansion known as cosmic inflation.

Galaxies evolve (clusters and superclusters form)	Solar system forms	Death of the Sun	Fate of the Universe
1 billion years	9 billion years	18.7 billion years	

COSMIC INFLATION AND THE CMB

It is thought that the temperature variations seen in the cosmic microwave background can be explained by something called quantum uncertainty.

Energy fluctuations in the quantum foam became imprinted on the Universe seed when it was still smaller than a proton.

As the Universe inflated, the fluctuations expanded along with it.

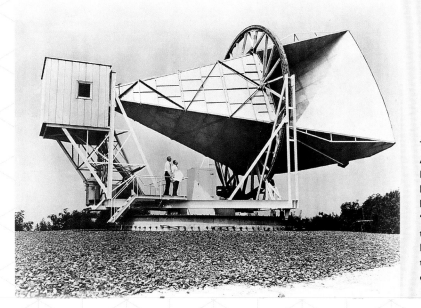

Quantum fluctuations appear in the Universe seed

Rapid inflation 'fixes' and expands the fluctuations

The Holmdel Horn Antenna or 'Large Horn Antenna' in New Jersey, used by Penzias and Wilson to 'accidentally' discover the Cosmic Microwave Background (CMB) – the radiation afterglow of the Big Bang.

The CMB was first detected by astronomers Arno Penzias and Robert Wilson in 1964 at the Large Horn antenna (a radio telescope that resembled a giant hearing aid horn). While testing the new telescope, they picked up a strange static-like hiss of noise, which they could neither explain nor eliminate. As they became increasingly irritated by the 'malfunction' they discovered that a family of pigeons had made their home in the antenna. Assuming the hiss was the result of pigeon poo, they scraped it off and (just to be sure) shot the pigeons. But the hiss stubbornly refused to go away.

The same year, the two physicists who had first suggested the existence of the cosmic microwave background, Robert Dicke and George Gamow, put two and poo together and realized that the errant hiss was the radio signal from the Big Bang. In 1978 Penzias and Wilson were awarded the Nobel Prize in Physics for the discovery of the CMB, which they had neither looked for nor recognized – Dicke and Gamow received nothing.

At first, the discovery of the CMB was seen as a great vindication of Big Bang theory, but it soon became apparent that there was a problem: it was just *too* perfect. The early measurements seemed to show that, no matter where you looked in the sky, the background radiation was perfectly uniform. But Big Bang theory predicted that, if the Universe had been born in a giant explosion, the distribution of matter in the early Universe should be lumpy (like debris thrown from an explosion) and, as a result, the background temperature should be lumpy also. A uniform CMB would have meant that the Big Bang fireball had also been perfectly uniform, which seemed impossible.

Worse still, if the radiation *were* perfectly uniform, the whole standard model would have fallen apart, since if there had been no irregularities (patches where matter was a little denser) there would have been no seeds from which stars and galaxies could take root, and you, dear reader, wouldn't exist.

Luckily, you do exist (as does the rest of the Universe) and, as instruments became more sensitive, it became clear the CMB wasn't quite as uniform as it had first appeared and that there are indeed tiny imperfections in the background radiation. But Big Bang wasn't out of the woods yet: it couldn't explain why the background radiation was so 'almost' perfect. That the temperature was so smooth seemed to suggest that somehow the Universe had violated one of the most unbreakable laws of physics – the speed of light.

GALAXIES OUT OF SIGHT

Here are two distant galaxies. Both are visible from Earth, but they cannot see each other. This is because the Universe is only 13.8 billion years old and light just hasn't had enough time to travel between them. If light hasn't had time to make the journey, then no other information has either. For example, because our Sun is 8.4 light minutes from Earth, if it were to suddenly vanish, we wouldn't feel the effects of its disappearance for 8.4 minutes.

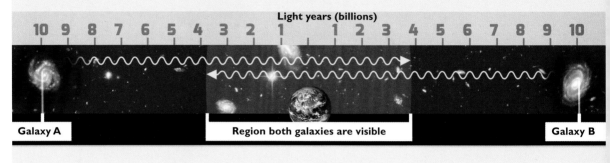

Light years (billions)

10 9 8 7 6 5 4 3 2 1 1 2 3 4 5 6 7 8 9 10

Galaxy A

Region both galaxies are visible

Galaxy B

The whole vast expanse of the Universe was pretty much the same in all directions, but how did regions of space divided by so much distance 'know' how to be the exact same temperature at the exact same time?

It's time to introduce a word the physicists like to use, but that you and I will probably never use in polite conversation: entropy. Entropy is a state of equality that nature strives to achieve. Basically it means that an organized system of energy is inherently unstable and, as such, will seek stability by becoming disorganized (a pencil stood on end is an organized system possessing potential energy and, as such, is unstable – it *will* 'decay' towards entropy by falling over and spending its energy).

The same is true of temperature differences. If you put a hot cup of coffee with a temperature of 70°C in a kitchen that is 20°C, the organized energy within the coffee will pass into the room until the temperature difference has been equalized and maximum entropy has been reached – this is what the second law of thermodynamics describes.

If, as Big Bang theory had first predicted, the Universe had been created all lumpy with hotter regions and colder regions, it would have behaved just like the coffee cup in the kitchen: energy from warm spots would have passed into cooler spots until the Universe was the same temperature from one end to the other.

But this is exactly what the CMB is telling us, so what's the problem?

The problem lies with the age of the Universe – there just hasn't been enough time for the energy to even out. Think about it this way. Look at two regions of space that are 10 billion light years away from us in opposite directions. Relative to each other, they are 20 billion light years apart and yet they will exhibit almost exactly the same properties: the same background temperature, the same sorts of galaxies distributed in much the same way.

It would take 20 billion years for light (and therefore energy) to travel the distance between them but, since the Universe is only 13.8 billion years old, it hasn't had enough time to make the journey.

The solution to this problem was cosmic inflation. Only when the Universe was very small would regions on opposite sides have been close enough to exchange heat. Also, the Universe must have started off expanding relatively slowly otherwise there wouldn't have been enough time for the temperature to even out.

However, if the Universe had continued to expand at this relatively relaxed pace, it wouldn't be anywhere near the size it is today. By waiting until the Universe has had time to 'even out' and then adding an extra 'kick' of energy (released by the breakdown of the unified force) to drive rapid expansion, we can have a Universe that is just the right size and just the right temperature.

But the CMB didn't stop its mischief-making there. As the sensitivity of our telescopes improved and we could make ever more accurate readings of the microwave background, another problem emerged: now the temperature wasn't uniform enough!

THE BIG BANG'S BRIEF TELEVISION CAREER

Before the advent of digital television, the screen of a de-tuned TV, or one tuned into a channel with a bad signal, would be graced with a dancing black and white pattern of hissing static. But what the cursing viewer was probably unaware of as he scrambled around trying to restore the big game to his screen, was that about one per cent of that static came from the CMB. Light from the Big Bang had travelled more than 13 billion light years and, stretched into radio waves, had been collected by the TV antenna and translated into TV fuzz – a rather ignominious end to such an epic journey.

ENTER THE QUANTUM FOAM

The most recent telescope to probe the cosmic microwave background is the European Space Agency's Planck space telescope. Named after German theoretical physicist Max Planck, the telescope measured the CMB in exquisite detail and, for the most part, confirmed Big Bang theory as the best explanation of the Universe's origins. It pushed back the age of the Universe from 13.73 billion years to 13.82 billion years and refined measurements of the amount of normal matter, dark matter and dark energy that make it up (more on these dark mysteries later). Planck won the Nobel Prize in 1918 for his work on quantum theory.

In 2013, the spacecraft created the most detailed map yet of the CMB and, in doing so, revealed that the most simple model of cosmic inflation doesn't quite fit. The Planck telescope confirmed something that its less sensitive predecessors had hinted at: that the Universe wasn't perfectly uniform after all and there are more temperature fluctuations than cosmic inflation predicted.

Luckily physicists don't have to dig too deeply into their kitbag of quantum weirdness to pull out a solution for this problem.

According to something called quantum uncertainty, what we think of as empty space is never truly empty. Were you able to peer beyond the realm of atoms and look into space at the very smallest quantum scale, you would reach something known as the quantum foam.

The quantum foam is the theoretical foundation of space-time (see Chapter 3 – Quantum

AND YOU THINK PARTICLES ARE TINY

To experience the quantum foam you would have to shrink yourself down to something called the Planck length. Named after the father of quantum theory, Max Planck, the Planck length is a sort of minimum size limit – nothing is smaller or shorter.

It is so tiny that, if you were to measure the diameter of an atom by pacing out one Planck length every second, it would take you ten million times longer than the Universe has existed (10,000,000 × 13,800,000,000 years) to complete the journey.

Weirdness, pages 64–65) – the weave that makes up the fabric of the Universe. At this scale, matter and energy can literally 'pop' into existence to create a bubbling foam of tiny particles that borrow energy from the Universe to pop up to say hello and then pop off again to repay their energy debt.

This game of particle whack-a-mole has little obvious impact on today's Universe but, back when it was smaller than an atom, these tiny variations were a very big deal indeed (like a drop of rain is a big deal to a flea).

Any quantum fluctuations present at the moment cosmic inflation began would have expanded along with the Universe and been imprinted on the CMB from that moment forwards (rather like painting dots on a deflated balloon and inflating it – the once tiny dots become much larger with cosmic inflation).

They might have been problematic for theoreticians but on the whole those temperature variations were a very good thing indeed. They reflect the differences in density and distribution of matter (more matter, more heat) that were imprinted on the Universe at the end of inflation and really are the blueprint for the cosmos that we'll build throughout the rest of the book. All those slightly denser regions had just a little more gravitational pull than their surroundings and, although it took some 400 million years, they would form the gravitational seeds around which matter would slowly accumulate to create the first stars and galaxies.

But, for now, our Universe is about to go dark.

Ironically, just as the stable atom was born from the cloud of super-heated plasma, and first light (or CMB) spread across the Universe, the cosmos was once again plunged into blackness. The flaming plasma cloud had been extinguished and, with no stars to send new light across the heavens, the period known to astronomers as the 'dark ages' had begun. The Universe will remain cold, dark and lifeless for the next few hundred million years.

While we're waiting around for things to kick off once more, let's take the opportunity to get to know the particle building blocks (and the fundamental forces that hold them together) that we will use to build our Universe.

The adventures of our developing Universe will continue in Chapter 4 – The Force is Strong with this Universe, page 70.

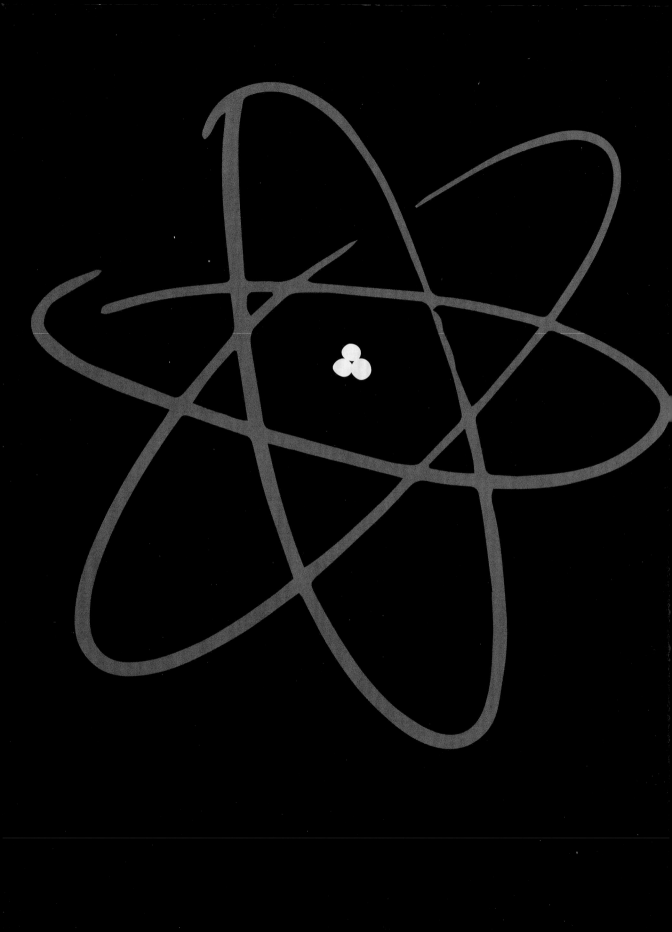

HOW WE DISCOVERED THE ATOM

In which we sit around the cosmic campfire and tell the tale of how we discovered the atomic building blocks of the Universe and how the atom was revealed to be far stranger than anyone could have imagined ...

So, the Universe is transparent and the photons carrying the radiation after-glow of the Big Bang (CMB) are beginning their almost 14-billion-year journey to reach our telescopes. There won't be any new photons being made for a few hundred million years so we've got a little time to kill until the lights are switched back on.

Before the invention of the television in the early 20th century, humans used to fill dark, lonely nights by telling stories to each other, so come gather around the fading embers of the Big Bang and I'll tell you the tale of how we discovered the atom ...

THE INDIVISIBLE ATOM

The story of the atom begins in the 5th century BC in Ancient Greece. A tunic-clad thinker called Democritus formulated the idea that matter might be comprised of tiny particles that were 'atomos', or indivisible. These 'atoms' couldn't be broken up, as there was nothing smaller for them to be broken into.

Then, for the next 2,000 years or so, nothing much happened and the atom was largely ignored. It made its next fleeting appearance on the European stage in the 17th century, when the Irish chemist, Robert Boyle, proved that gases are made up of widely-spaced atoms and it popped up again, in the 18th century, when the French chemist, Antoine Lavoisier, compiled the first list of the chemical elements.

In the early 19th century, an English physicist and chemist called John Dalton formulated his atomic theory. Dalton's atom was much the same as the ancient Greeks', but he went on to suggest the different elements were made of atoms of different sizes and that the elements could be combined to create more complex compounds. He was also the first person to make a serious attempt to calculate the atomic mass of some of the chemical elements and to introduce a system of chemical symbols.

A few decades later, in Russia, a chemistry teacher called Dmitri Mendeleev set about pondering the chemical elements. At that time, the elements tended to grouped either by their atomic weight or by what they reacted with, but Mendeleev believed they must possess some sort of underlying order. He spent more than 13 years collecting his own data and corresponding with scientists around the world. When he felt he had all the parts of his puzzle, he wrote the name of each element and its atomic weight on piece of card and tried to organize them. After spending three days and nights locked away playing his game of 'chemical solitaire', he had compiled a table that listed the elements by their atomic weight and grouped them into nine families (such as gases, metals, non-metals) and, in 1869, he published his Periodic Table of the Elements.

Mendeleev's periodic table revolutionized our understanding of the properties of atoms and lifted the curtain on the stage where the atom would truly receive the attention it deserved: the science of quantum mechanics.

The next big step took place in 1897. A British physicist, called Joseph John 'JJ' Thomson, was trying figure out the nature of cathode rays – a mysterious 'ray' emitted by a cathode (the negative part of an electrical conductor) within a vacuum tube. When he applied a positive charge, he noticed that the rays were attracted to it – meaning that they must carry a negative charge.

But the real breakthrough came when he calculated their mass and discovered they were about 1,800 times less massive than even the lightest atom (hydrogen). Since they were so small, he concluded that they must have come from inside atoms: the indivisible atom had been revealed to be divisible.

THE BILLIARD BALL ATOM

5TH CENTURY BC ...

Democritus (460BC–370BC)

Democritus believed the atom was a solid and indivisible sphere.

...TO EARLY 19TH CENTURY AD

John Dalton (1766–1844)

Oxygen + Hydrogen + Hydrogen

= Water

Dalton kept the 'billiard ball' model, but proposed that different elements had atoms of different sizes. He also suggested that different elements could be combined to create chemical compounds.

Thomson called these tiny negatively-charged particles 'electrons' and incorporated them into a revolutionary new model of the atom. He knew that atoms are neutral (carrying no overall electrical charge) so, to balance out the negative electrons, he imagined the atom being a sort of cloud of positive charge that was peppered with electrons – like bits of plum in a plum pudding.

Although Thomson went on to win the Nobel Prize in Physics for his discovery of the electron, his plum pudding model of the atom would only last about ten years.

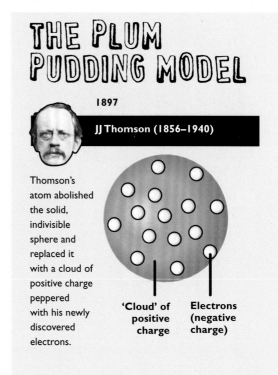

THE PLUM PUDDING MODEL

1897

JJ Thomson (1856–1940)

Thomson's atom abolished the solid, indivisible sphere and replaced it with a cloud of positive charge peppered with his newly discovered electrons.

'Cloud' of positive charge

Electrons (negative charge)

A SOLAR SYSTEM IN MINIATURE

In 1909, a New Zealand-born physicist, Ernest Rutherford, was looking over the results of an experiment performed by two of his students when he spotted a flaw in Thomson's atomic model. The two students, Hans Geiger and Ernest Marsden, were experimenting with radiation by firing positively charged particles at a piece of gold foil. Based on Thomson's model of the atom, they had expected the particles to shoot virtually unimpeded through the positive 'cloud' of the atom – which, although positively charged, should have been diffuse enough to allow the heavier particles to barge through it.

Instead, they saw that, while many of the particles did pass through, some were deflected and a very small number of the others bounced right back.

This led Rutherford to conclude that – instead of the positive charge coming from pudding containing electron plums – the atom must possess an extremely localized concentration of positive charge at its centre. He proposed that the nucleus was made up of distinct units of matter that he called protons and he placed Thomson's electrons into scattered orbits around the nucleus like planets orbiting the Sun.

Under Rutherford's new planetary model, the atom was revealed to be made up of almost entirely empty space, with most of its mass concentrated in the tiny nucleus. But he had a problem: what was stopping the negatively charged electrons being pulled into the positively charged nucleus?

To get around this, Rutherford dug into the kit bag of classical Newtonian physics and suggested that, just as the planets are kept in orbit by being accelerated by the Sun's gravity, electrons must be undergoing constant acceleration as they whip around the nucleus, which stops them falling out of orbit.

Unfortunately, as would become increasingly clear in coming years, although old-fashioned Newtonian physics works beautifully in the macro world, it just doesn't cut the mustard in the quantum world.

HOW BIG IS A NUCLEUS?

If you were to scale an atom up to the size of the Earth, the nucleus would only be about the size of a football stadium: the rest of the atom is empty.

This is where one of the founding fathers of quantum theory, Niels Bohr, enters the story. Bohr, a Danish physicist (and former footballer), saw the quantum flaw in Rutherford's otherwise ingenious atomic model. He thought back to the work the brilliant Scottish physicist, James Clerk Maxwell, had done on electromagnetism in the previous century. Maxwell had shown that, when an electric charge is under acceleration, it loses energy by emitting radiation (a process exploited by X-ray machines).

THE PLANETARY MODEL

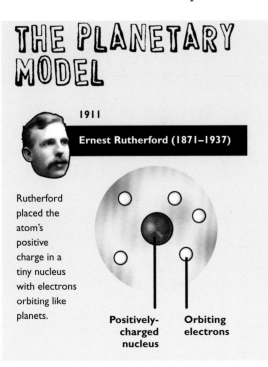

1911

Ernest Rutherford (1871–1937)

Rutherford placed the atom's positive charge in a tiny nucleus with electrons orbiting like planets.

Positively-charged nucleus

Orbiting electrons

BOHR'S ELECTRON SHELLS

1913

Niels Bohr (1885–1962)

Bohr realized that Rutherford's orbiting electrons would fall into the nucleus ...

... so he locked the electrons into fixed orbits based on their energy.

Electron in high-energy orbit

Electron in low-energy orbit

Bohr realized that Rutherford's accelerating electrons should lose energy by the same process and quickly fall into the nucleus – since this doesn't happen something else must be keeping an atom's electrons in check.

On 6 March 1913 (see 'And you think Particles are tiny', page 50) Bohr explained his modifications to the planetary model in a letter to Rutherford. Building on the work done by Max Planck, who had shown in 1899 that there was a limit to how far something could move, or be divided, at the quantum level (there is a smallest quantum distance that cannot be divided – called the Planck distance), Bohr proposed that electrons are restricted to fixed orbits depending on their energy.

Electrons with the least energy occupy the lowest orbit (which can't get any lower) and those with the most energy occupy the highest orbits – they can only move between these orbits, or shells, by gaining or losing energy.

This also answered another mystery of the atom. Scientists had noticed that, when an atom was heated, it emitted radiation in very specific amounts and no one had been able to explain this. Bohr's model did, however, by saying that, when the atom is heated, its electrons gain energy, which gets them all excited and causes them to 'leap' into a higher energy orbit (the origin of the term quantum leap). When the excited electron calms down, it emits a packet of energy (in form of a photon) and drops back into its original lower orbit.

So the electrons had been put in the right place at last, but there was still something missing from the atomic model – something literally didn't add up. Rutherford noticed that, in most cases, the atomic number (number of protons) of a chemical element was only around half that of its total atomic weight – it was as if something extra was 'hiding' inside the atom.

In 1920, he suggested that the extra something might be an as-yet-unseen particle that had about the same mass as the proton but, rather than being electrically-charged, would possess no charge at all – a neutral particle that wouldn't upset the balance between the positive proton and the negative electrons. Rutherford called his hypothetical particle the neutron.

The hunt for the neutron was on and the man to find it would be Rutherford's assistant, the British physicist James Chadwick.

Being neutrally-charged, the neutron was rather difficult to locate. Fortunately, discoveries in Europe would provide just the trail of breadcrumbs that Chadwick would need to track the neutron down.

In 1930, researchers in Germany discovered that if you bombard the element beryllium with alpha particles (a particle we now know to consist of two protons and two neutrons – like a helium atom but without the electrons), a strange neutral radiation was emitted that

PERIODIC TABLE TROUBLES

A quick glance at the periodic table reveals that the atomic number of many elements is less than half of its atomic weight.

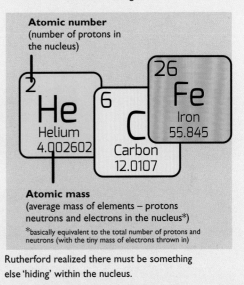

Atomic number
(number of protons in the nucleus)

Atomic mass
(average mass of elements – protons neutrons and electrons in the nucleus*)

*basically equivalent to the total number of protons and neutrons (with the tiny mass of electrons thrown in)

Rutherford realized there must be something else 'hiding' within the nucleus.

DIVIDING THE NUCLEUS

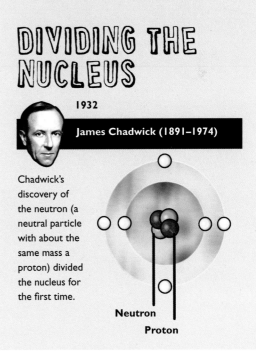

1932

James Chadwick (1891–1974)

Chadwick's discovery of the neutron (a neutral particle with about the same mass a proton) divided the nucleus for the first time.

Neutron

Proton

could penetrate matter. Chadwick became convinced that this neutral radiation was Rutherford's missing neutral particle.

Experiments in France had shown that, when paraffin wax was placed in its path, the neutral radiation could 'knock' protons from the paraffin's atoms. To Chadwick, this was proof that a particle was at play.

He reasoned that only a particle could knock other particles from an atom. Imagine the particles within the paraffin are like a pack of snooker balls. When you fire a cue ball at the pack, some of the balls are knocked out and scatter away from the impact – just like the protons are knocked from the paraffin.

Chadwick replicated the paraffin experiment and he not only confirmed that the neutral radiation was indeed a particle but also, by tracing the paths and energies of the dislodged protons, was able to figure out that the particle must have about the same mass as the protons it had dislodged.

So, after centuries of detective work, scientists finally had an accurate model of the atom to work with. The once indivisible sphere of matter had been divided into a dynamic system of protons, neutrons and electrons; but it didn't end there.

ELEMENTARY MY DEAR WATSON

In the following decades, the atom would be further divided into a veritable pantheon of particle building blocks. There were quarks: the up quark and down quark – protons and neutrons are made of different combinations of these quarks – and several more massive relatives with funky names like the charm and strange quarks.

Although it turned out that the electron wasn't made of anything smaller, it would be joined by its own heavyweight cousins – the muon and tau. Each of the fundamental forces (electromagnetism and the strong and weak nuclear forces) would get their own particle representatives.

Then physicists discovered that each particle had its own antiparticle – an antimatter twin identical in mass but opposite in everything else.

By the mid-1930s, it seemed that physics had reached the limits of the indivisible's divisibility. They had dissected the atom into its smallest components – called the elementary, or fundamental, particles – and, around these building blocks, physicists constructed a theoretical framework to describe how they work and interact together, called the Standard Model of particle physics.

Despite all the additions, the basic model of the atom has barely changed. Even today, more than a century after it was first devised, Rutherford and Bohr's model is still taught in classrooms around the world and is used in countless science books (such as this). It has an elegant simplicity that almost seems too good to be true – which is hardly surprising, because that is exactly what it is.

ATOMS IN WONDERLAND

We are attracted to the classroom model of the atom because it is something we can relate to. It is a sphere made of smaller spheres, orbited by much smaller spheres locked into spherical shells. We can easily imagine those spheres bonding together and we can easily picture them being knocked around and dislodged like teeny tiny billiard balls. But really, it's just a metaphor – a stylized image designed to make the atom seem to fit the reality of our day-to-day existence. In reality, the atom, as you know it, doesn't exist.

The tale of the atom's discovery is also the tale of how a whole new branch of science was born; one that was so weird and counterintuitive that it would confound and divide the very scientists who created it: quantum mechanics. (The following pages of graphics illustrate some of its mindboggling features.)

The story of how quantum mechanics came to be (and the world it revealed) is too convoluted

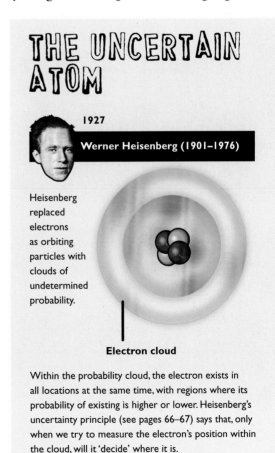

THE UNCERTAIN ATOM

1927

Werner Heisenberg (1901–1976)

Heisenberg replaced electrons as orbiting particles with clouds of undetermined probability.

Electron cloud

Within the probability cloud, the electron exists in all locations at the same time, with regions where its probability of existing is higher or lower. Heisenberg's uncertainty principle (see pages 66–67) says that, only when we try to measure the electron's position within the cloud, will it 'decide' where it is.

CONTINUES ON PAGE 68 ➡

PARTICLE BUILDING BLOCKS
THE SMALL STUFF THAT BUILDS PARTICLES

According to the Standard Model of particle physics, atoms are made of particles, which in turn are made of elementary particles.

So, let's turn the elementary particles into some building blocks we are a little more familiar with: Lego (other modular construction toys are available).

This is the atom as we have got to know it so far:

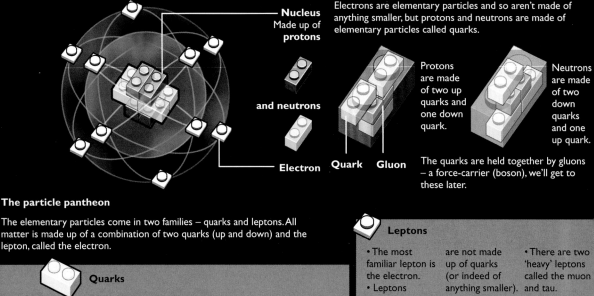

Nucleus
Made up of
protons

and neutrons

Electron   Quark   Gluon

Electrons are elementary particles and so aren't made of anything smaller, but protons and neutrons are made of elementary particles called quarks.

Protons are made of two up quarks and one down quark.

Neutrons are made of two down quarks and one up quark.

The quarks are held together by gluons – a force-carrier (boson), we'll get to these later.

The particle pantheon

The elementary particles come in two families – quarks and leptons. All matter is made up of a combination of two quarks (up and down) and the lepton, called the electron.

Quarks

• All of the matter in the universe is made of a combination of up and down quarks.
• All particles composed of quarks are called hadrons (Greek for heavy).
• Protons and neutrons are also known as baryons.
• Quarks come is six 'flavours', which have different properties and masses.

Leptons

• The most familiar lepton is the electron.
• Leptons are not made up of quarks (or indeed of anything smaller).
• There are two 'heavy' leptons called the muon and tau.

• Another lepton is the neutrino – a ghostly, almost massless particle that hardly interacts with matter.

Bosons (force carriers)

• Bosons are particle messengers that tell other particles how to interact with the fundamental forces (the strong force, weak force, electromagnetism and gravity).

Gluon
• This mediates the strong nuclear force and is responsible for holding quarks together to form protons and neutrons.

Photon
• This tiny package of energy carries the electromagnetic force, which affects any fundamental particle that carries a charge.

W & Z bosons
• These mediate the weak nuclear force, which is responsible for radioactive decay.

Higgs boson
• The particle representative of the Higgs Field, which gives mass to quarks and leptons.

Graviton
• A theoretical particle that (if it exists) is responsible for carrying the gravitational force.

We'll be looking at the fundamental forces and their force carriers in the next chapter.

MATTER

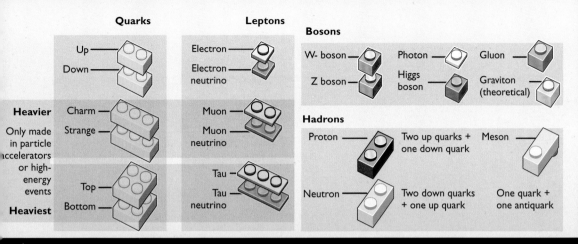

Quarks
- Up
- Down
- Charm
- Strange
- Top
- Bottom

Leptons
- Electron
- Electron neutrino
- Muon
- Muon neutrino
- Tau
- Tau neutrino

Heavier

Only made in particle accelerators or high-energy events

Heaviest

Bosons
- W- boson
- Z boson
- Photon
- Higgs boson
- Gluon
- Graviton (theoretical)

Hadrons
- Proton — Two up quarks + one down quark
- Neutron — Two down quarks + one up quark
- Meson — One quark + one antiquark

Each Standard Model particle has an antimatter equivalent in which the particle's properties (charge and spin) are reversed – a positive charge becomes negative; a negative becomes positive; a neutral particle remains neutral but other properties are reversed.

ANTIMATTER

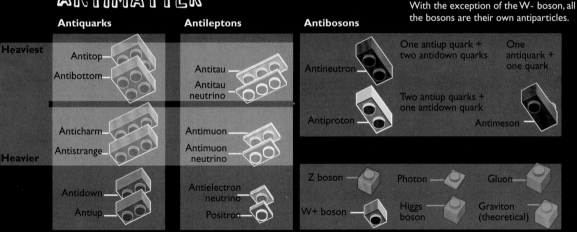

With the exception of the W- boson, all the bosons are their own antiparticles.

Antiquarks
- Antitop
- Antibottom
- Anticharm
- Antistrange
- Antidown
- Antiup

Heaviest

Heavier

Antileptons
- Antitau
- Antitau neutrino
- Antimuon
- Antimuon neutrino
- Antielectron neutrino
- Positron

Antibosons
- Antineutron — One antiup quark + two antidown quarks
- Antiproton — Two antiup quarks + one antidown quark
- Antimeson — One antiquark + one quark
- Z boson
- Photon
- Gluon
- W+ boson
- Higgs boson
- Graviton (theoretical)

NOMENCLATURE NONSENSE?

The names given to particles can be confusing:

- Elementary particles are sometimes referred to as fundamental particles.
- Particles made of quarks are collectively known as hadrons but they are also known as baryons.
- But not all hadrons are baryons – hadrons made of one quark and one antiquark are called mesons and these are not baryons.
- Quarks and leptons are collectively known as fermions.
- Fermions can also include hadrons like the proton.

Luckily, this doesn't really matter to us – we will be mostly playing with protons, neutrons, electrons and photons and, to a lesser degree, some quarks and a little antimatter.

TIP OF THE ICEBERG

There are a lot of particles on this page to get to grips with, but these 'standard' particles of the Standard Model could be just the tip of the iceberg.

Supersymmetry is an extension to the Standard Model in which every fundamental particle has a 'twin' (or multiple partner) particle. If they exist, these super-particles (or sparticles) will have much more mass than their Standard Model cousins.

QUANTUM WEIRDNESS
HOW A PARTICLE CAN BE A WAVE AND A PARTICLE

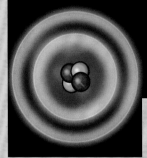

The world of quantum mechanics is baffling, bizarre and just plain weird. It goes against everything that our macroscopic existence teaches us to expect and much of it has the potential to transform the human brain into a quivering mess of gelatinous denial. So, to begin the mental jelly transformation, I give you wave-particle duality...

When we study physics at school, we are taught that particles are like teeny-tiny marbles – small, spherical packets of matter that make up the world around us. But the truth is far weirder than that. Let's start with a famous experiment:

THE DOUBLE SLIT EXPERIMENT

1 This is what happens when you shoot light particles (photons) through a slit in a piece of paper.

Single slit

Particle

Single lines

As you'd expect, they pass through the slit and leave behind a single vertical mark on a detector at the back – just as if you'd fired a bunch of marbles through it.

2 So, what happens if we add another second slit? You'd expect the particles to leave two matching lines on the sheet at the back, right? But look what actually happens ...

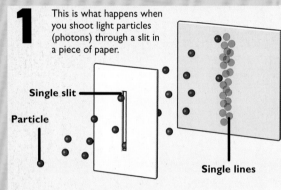

Double slit

Particles

Multiple lines

Instead of two lines (as your macroscopic mind would expect), there are many lines. How is this possible?

3 The only explanation for this is that the particles are behaving as if they are a wave.

When a wave passes through two slits, it spreads out in two fronts which overlap each other.

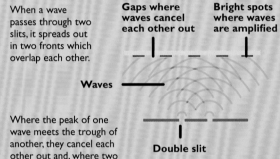

Gaps where waves cancel each other out

Bright spots where waves are amplified

Waves

Where the peak of one wave meets the trough of another, they cancel each other out and, where two peaks meet, they amplify each other. This creates an interference pattern on the sheet at the back – identical to the lines left by the particles.

Double slit

4 Weirder still, if you fire just one particle at a time through the slits, you still get an interference pattern.

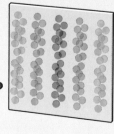

This means that each single photon must have passed through both slits at the same time and then interfered with itself as a wave.

Niels Bohr

(received the Nobel Prize for his contribution to quantum mechanics)

Erwin Schrödinger

(received the Nobel Prize for his contribution to quantum mechanics)

Richard Feynman

(received the Nobel Prize for his contribution to quantum electrodynamics)

'If anybody says he can think about quantum physics without getting giddy, that only shows

he has not understood the first thing about them.'

'I do not like it, and I am sorry I ever had anything to do with it.'

'I think it is safe to say that no one understands quantum mechanics.'

5 If that wasn't weird enough for you, let's install a detector at the slits, which beeps every time a particle passes through one of the slits.

Beep!

Particle detector

It will see each particle pass through just one slit and the detector at the back will have just two lines on it. It's the same test as before, but this time there is no interference pattern.

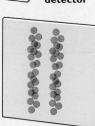

In other words, when we test for wave-like properties, the photon behaves as a wave. But, when we test for particle-like properties, it behaves like a particle.

It seems that, when left to its own devices, the photon exists as both a wave and a particle. The act of observing the particle forces it 'choose' what it will be!

The particle as described by quantum mechanics is like an unsupervised child – doing whatever it wants, wherever it wants – right up to the moment a parent (or other macroscopic observer) catches it in the act.

WAVE-PARTICLE DUALITY

A single particle is able to pass through two slits at the same time because, at the quantum level, matter doesn't really exist at a fixed state. Instead it exists in a cloud of 'probability' called the wave function.

The wave function was thought up by German physicist, Erwin Schrödinger (he of dead/alive cat in a box fame). It basically means, instead of thinking of it as defined point of mass, a particle is more of a region of wavy potential smeared across space.

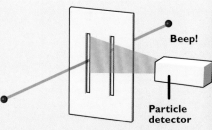

Wave function

Direction of travel

Like all waves, a particle wave is spread over a large area, so it has no definite position (peaks and troughs are regions where the probability of the particle occurring increase), but it does have a direction in which it is travelling. As a wave, we can know its direction of travel, but we can't know the position of the particle.

Likewise, if we measure the position of the particle, the wave function collapses and we can no longer measure its direction of travel. This inability to measure all a particle's properties is called Heisenberg's Uncertainty Principle.

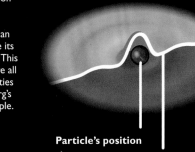

Particle's position

All other probabilities vanish and wave function collapses

VIRTUAL PARTICLES

HOW TIME-TRAVELLING ELECTRONS CAN MAKE SOMETHING FROM NOTHING

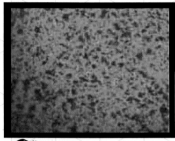

On pages 50–51, we talked about something called the quantum foam where 'matter and energy can literally "pop" into existence' as if from nowhere. This is how it happens ...

If you thought waves that are particles was weird, prepare for a new level of brain-melting quantum goodness. It comes to us from the brilliant mind of American physicist (and bongo player) Richard Feynman (with a little help from Albert Einstein and Werner Heisenberg).

1 First up, Einstein

Relativity tells us that space and time are inextricably connected (space-time) and the faster you travel through space, the slower you move through time – someone travelling at near-light speed will experience less time passing than a stationary observer (relative to the traveller).

In theory, if that person could travel faster than light, they would appear (once again to that observer) to be moving backwards in time (which is why light speed is considered an unbreakable cosmic speed limit).

2 Over to Heisenberg

As we've seen, Heisenberg's Uncertainty Principle tells us that a particle exists in all states, doing all things at the same time – only by measuring it do we constrain its actions and force it to assume a single property.

Basically it says that 'if you can't see it happening, anything is possible'. This is as true of 'empty' space as it is of the particle.

The shorter the amount of time you look at something, the less certain you can be of what is going on.

In quantum physics, there is a shortest period of measurable time called the Planck time – anything that happens within that time is, by definition, unmeasurable and, if it's unmeasurable, Uncertainty tells us that anything is possible.

3 Feynman's time-travelling electron

Richard Feynman imagined an electron whizzing around through space. In the middle of its journey the electron speeds up to faster-than-light speed. Rather handily, this one has decided to move around on a set of axes (not the tree-chopping kind) depicting space and time.

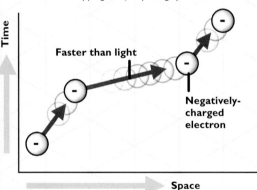

Time / Faster than light / Negatively-charged electron / Space

4

He realized that relativity tells us that to another observer, the electron would appear to be moving backwards in time during its faster-than-light period – so he would see the electron moving forwards in time, then backwards and then forwards again.

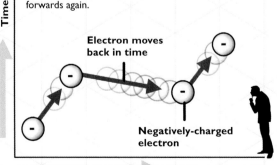

Time / Electron moves back in time / Negatively-charged electron / Space

5 To a physicist like Feynman, a negative charge moving backwards in time is equivalent to a positive charge moving forwards in time. The reversal of time causes the properties of the electron to become reversed also – and a negatively-charged electron with its properties reversed is a positron (the electron's antimatter equivalent).

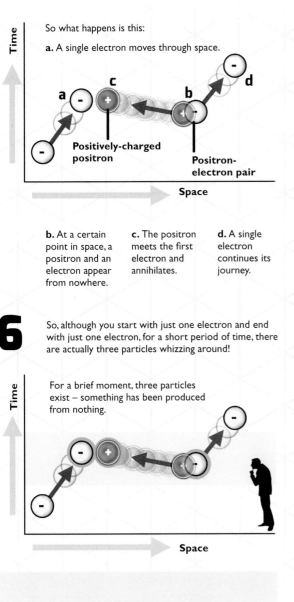

So what happens is this:

a. A single electron moves through space.

Positively-charged positron

Positron-electron pair

Space

b. At a certain point in space, a positron and an electron appear from nowhere.

c. The positron meets the first electron and annihilates.

d. A single electron continues its journey.

6 So, although you start with just one electron and end with just one electron, for a short period of time, there are actually three particles whizzing around!

For a brief moment, three particles exist – something has been produced from nothing.

Space

As long as the period of time between positron creation and annihilation is so short that we can't measure the particles (the Planck time where anything is possible) no rules have been broken – the particle can exceed the light speed limit and matter can appear from nowhere.

Particles that appear in this way are called virtual particles.

VIRTUAL REALITY

On the face of it, predicting the spontaneous appearance of something that, by definition, you can't hope to measure directly, seems as pointless as imagining that an unopened cupboard might contain invisible elephants, but physicists have detected virtual particles indirectly by looking for their effects on more measurable stuff.

1 This is a hydrogen atom consisting of one proton and one electron.

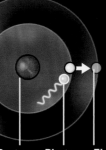

When an electron is struck by a light photon, it absorbs its energy, gets all excited and jumps into a higher-energy orbit.

When it calms down, the electron emits another photon and drops back to its original orbit.

Proton   Photon   Electron

It emits the light at a fixed set of frequencies that can be measured as light spectra.

2 Physicists have equations that can predict the frequencies of the absorbed and emitted light, but they don't always work. Sometimes the measured frequencies differ slightly from the predictions.

3 But this can be fixed by adding those virtual particles to the model.

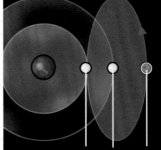

When physicists add a short-lived virtual electron and positron to the system, the spectra predictions agree to an accuracy of one part in a billion – making it the most accurate prediction in all of science.

Virtual electron   Electron   Virtual positron

This means that, at any given time, it is more accurate to show an atomic model with virtual particles than without them.

However bizarre it might sound, virtual particles DO exist and (as we'll find out) they actually make up most of the mass of an atom. Therefore, most of YOUR mass (and the Universe) is made of virtual particles!

and complex to squeeze into this book, but it is a story that involves almost every great scientific mind of the 20th century. Its list of founders reads like a who's who of modern physics: Max Planck, Albert Einstein, Niels Bohr, Erwin Schrödinger, Werner Heisenberg, Wolfgang Pauli, Paul Dirac, Louis de Broglie, Enrico Fermi and Richard Feynman (to name but a few).

Their discoveries would scoop Nobel Prize after Nobel Prize and make many of them household names and global celebrities. Yet the world they uncovered was so bizarre and counterintuitive that it would lead one of its founding fathers, Erwin Schrödinger, to declare 'I do not like it, and I'm sorry I ever had anything to do with it'.

Simplified greatly, in the world of quantum mechanics, matter exists in a state of constant flux – particles are transformed from neat, tangible spheres into diffuse clouds of probability in which they exist (and don't exist) in all positions (and none at all) and in all states (both as particles and waves) at the same time. It is a world where electrons can travel back in time; where matter can 'borrow' energy from the Universe and appear from nowhere; and where we can force particles to 'decide' what they will be (or whether they exist at all) just by trying to look at them. And yes, that's the simple version.

What this means for our atom is that, instead of neatly orbiting electrons, it now has clouds of probability – within which, the electron could be anywhere (and everywhere and nowhere).

The quantum world is undoubtedly a very strange place, yet it has made so many accurate predictions that (along with Big Bang theory and the Standard Model) it is considered one of the most successful theories in science – and it does have an important part to play in our story of the Universe.

But before we can put all those particle building blocks and their quantum quirks to use, we need something to hold everything together – we need some particle glue. We need some fundamental forces.

HOW CAN MATTER FORM FROM ENERGY?

●○

Einstein's famous equation, $E=mc^2$ (where 'E' is energy, 'm' is mass and 'c' is the speed of light) describes how mass and energy are really just different facets of the same thing. Mass and energy are interchangeable – mass can be converted to energy (as it is in the nuclear furnaces that power the stars) and energy can be converted into mass.

You can think of particles of matter as being tiny packets of super-concentrated energy – the more massive the particle is, the more energy it has locked away inside it (here 'massive' means that it has more mass, or matter, and not that it is necessarily large). Given the right amount of energy and pressure, you can literally squeeze energy until matter is created.

Before inflation, the Universe was just too hot and dense for matter to form but, for the briefest of moments after inflation, conditions were just right – the pressure had been released just enough for matter to coalesce (rather like bubbles can only form in a fizzy drink when the cap is removed and the pressure is relieved).

It might seem unlikely that matter of the sort that makes up you and me could be 'summoned' out of nowhere like some sort of particulate genie, but it really does happen. When physicists smash atoms together in particle accelerators like the Large Hadron Collider, they aren't just looking for bits of broken atom kicked out by the impact (like shards of glass thrown from a wrecked car). They are also looking for new particles that have been created from the intense pressure and energy found in the dense subatomic fireballs (a million times hotter than the centre of the Sun) formed at the point of collision. This is why you will hear scientists talking about the LHC 'recreating conditions at the time of the Big Bang'. It's not rhetoric – they really do create mini-Big Bangs in which the colliding protons melt into the same sort of hot and dense fundamental particle soup from which the Universe emerged.

This is also how we know so much about the first milliseconds of the Universe … it's been recreated in the lab.

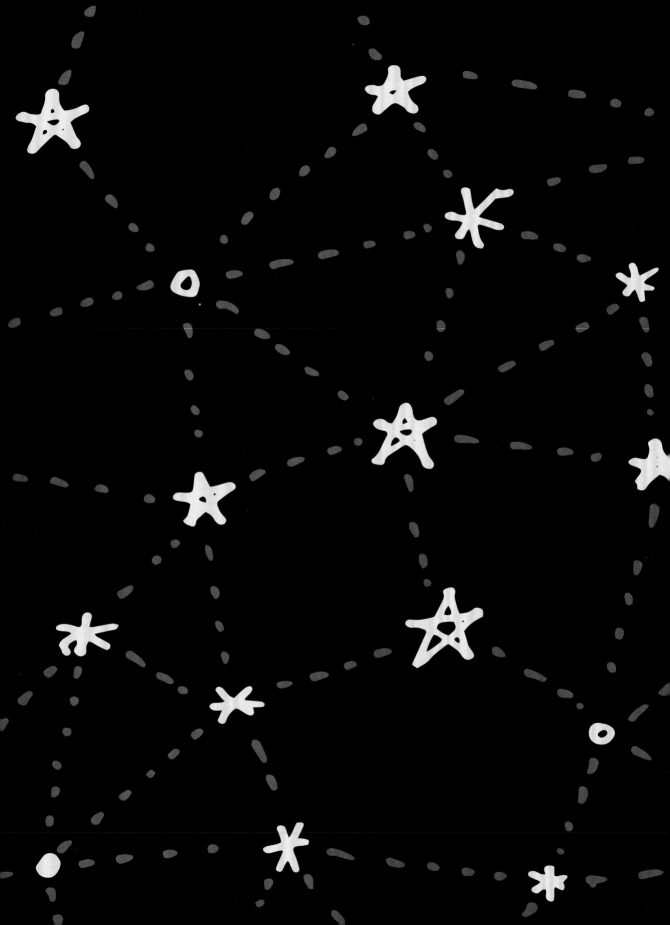

THE FORCE IS STRONG WITH THIS UNIVERSE

In which we explore the fundamental forces that underlie the workings of the Universe and, in the process, mix pop culture metaphors in a giant metaphorical metaphor mixer.

I n the *Star Wars* movies, Obi-Wan Kenobi describes the mystical force that gives the Jedi their powers as 'an energy field created by all living things. It surrounds us, penetrates us, and binds the galaxy together.'

In physics a force is less mystical but (on the face of it at least) no less mysterious and, in many ways, the Jedi definition fits quite nicely. The fundamental forces of the Universe do indeed surround us, penetrate us and they really do bind the galaxy (and the Universe together).

Physicists call them the fundamental forces because they are literally fundamental to the existence of the Universe. If the Universe was a house, the fundamental forces would be the foundations, the mortar and the beams that keep the bricks bound together and the walls supported – in fact, without the fundamental forces, we wouldn't even have the bricks with which to build the house.

In the *Star Wars* Universe, the force might have permeated all things but it needed a corporeal vessel (the Jedi) to channel it and deliver its powers (never call Vader a corporeal vessel ... he gets antsy). In the world of the Standard Model, the vessels used to deliver the power of the forces are particles called bosons.

These force carriers are exchanged between objects (such as quarks, protons, magnets, and even planets) when the relevant force is transmitted between them.

There are four fundamental forces (not counting the Higgs field, which is discussed in greater detail on page 78): the strong nuclear force, the weak nuclear force, the electromagnetic force and gravity – each of which (apart from gravity) has its own force carrier.

THE STRONG NUCLEAR FORCE

The strong force, or strong interaction, is responsible for holding together the stuff that forms the atomic nuclei. The nucleus of an atom is made up of protons, which carry a positive electric charge, and neutrons, which carry no electric charge. Left to their own devices, the particles that make up the atomic nuclei wouldn't stay together. The neutrons are OK. Being neutral, they have very little to say about things and

THE STRONG NUCLEAR FORCE

The strong force holds atoms together through the exchange of its force carrier, the appropriately-named gluon.

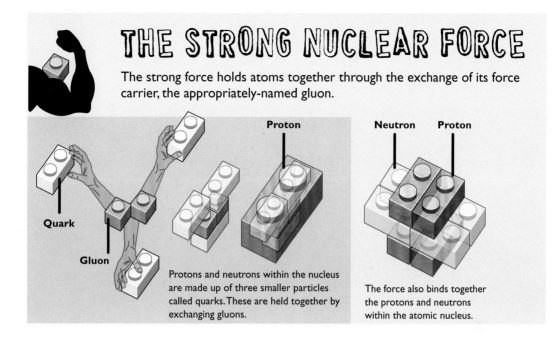

Quark

Gluon

Proton

Protons and neutrons within the nucleus are made up of three smaller particles called quarks. These are held together by exchanging gluons.

Neutron   **Proton**

The force also binds together the protons and neutrons within the atomic nucleus.

are neither enthusiastic about nor opposed to being in an enclosed space with other particles. The trouble lies with the protons.

Because protons carry a positive electric charge, they really object to being in close proximity to other protons – they literally find each other repellent and want to be as far as possible away from one another.

If you have ever played around with magnets, you will understand how much protons don't want to be close to each other. If you haven't (although I have to ask what on Earth you did with your childhood), take two magnets and try to push their positive poles together ...

... go on then ...

... try harder ...

... give up?

Yup, it's virtually impossible – they want to come together about as much as Yoda wants to snuggle down on the sofa and share a slanket and a cup of cocoa with Darth Vader.

The force with which two positively charged protons repel each other is almost insurmountable but, luckily for us, the strong nuclear force can do just that. It can overcome the protons' natural repulsion and bind them together within the atomic nucleus.

THE WEAK NUCLEAR FORCE

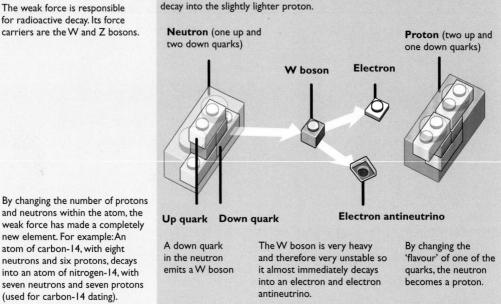

The weak force is responsible for radioactive decay. Its force carriers are the W and Z bosons.

The weak force can cause the slightly heavier neutron within an atom to decay into the slightly lighter proton.

Neutron (one up and two down quarks)

W boson

Electron

Proton (two up and one down quarks)

Up quark   Down quark

Electron antineutrino

By changing the number of protons and neutrons within the atom, the weak force has made a completely new element. For example: An atom of carbon-14, with eight neutrons and six protons, decays into an atom of nitrogen-14, with seven neutrons and seven protons (used for carbon-14 dating).

A down quark in the neutron emits a W boson

The W boson is very heavy and therefore very unstable so it almost immediately decays into an electron and electron antineutrino.

By changing the 'flavour' of one of the quarks, the neutron becomes a proton.

It's true to say that, without the strong nuclear force, there would be no atoms – and this would be a very short book, written on non-existent paper by a man who doesn't exist in a Universe that never was.

But although it is immensely powerful, the strong nuclear force only works over a very (very, very, very) short distance. If the two protons are separated by more than a proton's width, they fall beyond the reach of the strong force and can be influenced by the other forces (most notably the electromagnetic force, which has wanted to separate the protons all along). More importantly, the protons have to be able to get close enough to each other to be 'grabbed' by the strong force in the first place, which means overcoming their natural repulsion by some other means – this will become very important later on in the book when we build our Universe.

Rather aptly for a force that glues particles together, the strong interaction's force carrier (or boson) is the gluon.

THE WEAK NUCLEAR FORCE

Despite its rather derogatory name, the weak force is no less important than the strong force. The weak force is what makes the Sun (and all the stars) burn and is responsible for radioactive decay – allowing atoms to change their very nature by taking on, or losing (radiating), particles.

It is called the weak force because it is much weaker than the strong force and the electromagnetic force. Despite its relative feebleness, the weak force can have a dramatic effect over short distances (even shorter than the strong force).

In radioactive elements it is strong enough to break the bonds that hold the nuclei together. At its most simple, it can cause neutrons within the nucleus to decay into protons by making them lose an electron (or a positron). In doing so, it allows an atom to change into a different element.

In radioactive decay, the atom loses energy and (since energy and mass are interchangeable) it becomes a lighter element (one that has less mass).

But the weak force is also what makes nuclear fusion possible – the mechanism that powers the Sun – by allowing light elements to fuse together to create heavier elements and some spare energy. Without the weak force, the nuclear furnaces that power the stars would never have flared into life and the Universe would be a dark place indeed.

The weak interaction has two force carriers: the W and Z bosons, which are heavy particles with about 100 times the mass of a proton.

THE ELECTROMAGNETIC FORCE

The strong and weak nuclear forces act on an atomic level and so, on the face of it at least, have very little impact on our day-to-day lives. Now we start talking about the forces you can actually see and interact with.

The electromagnetic force works on any particle that carries an electric charge (so it has no effect on a neutron, for example) and its importance to our daily lives is

THE ELECTROMAGNETIC FORCE

The electromagnetic force affects any fundamental particle that carries a charge. Its force carrier is the photon.

1. Electron wranglers

Electrons are kept in orbit through the constant exchange of photons between protons and electrons. But, unlike the photons that we perceive as light, these are virtual photons whose effects can be seen but which can't be directly detected.

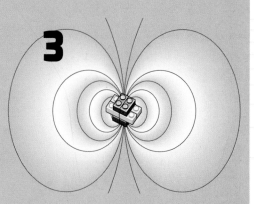

Electron (negative)

1

Virtual photon

Nucleus (positive)

2. Light trippers

Microwave photon

Infrared photon

X-ray photon

The whole of the electromagnetic spectrum is carried by photons (real photons, not virtual ones). Photons at the infrared end of the spectrum are less energetic (and have a longer wavelength) than photons at the X-ray end of the spectrum.

2

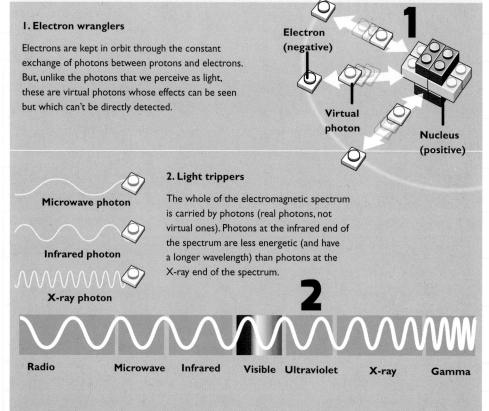

| Radio | Microwave | Infrared | Visible | Ultraviolet | X-ray | Gamma |

3. Magnetic personalities

The exchange of virtual photons is also responsible for the magnetic attraction and repulsion felt between two opposing magnetic poles and for the creation of magnetic fields.

The strange thing is (given how familiar we are with magnets) no one really knows how this works on a quantum level. By definition, you can't study the virtual photons (remember they exist for less than the Planck time) so figuring out how they work is rather tricky.

3

hard to overestimate. It is responsible for binding protons to electrons within atoms and it binds atoms together to create complex molecules.

The electromagnetic force is what gives the Universe its structure and shape. An atom is 99.999999 per cent empty space, so, without the electromagnetic force to give matter its structure, there would be nothing to stop atoms passing through each other. It is the interaction of all those electromagnetic attractions and repulsions that prevent you passing through the chair you are sitting in (or floor you are standing on). Think of it as being built of billions of tiny magnets with billions of tiny repelling poles that stop you merging with your chair and billions of attracting poles that stop you falling apart and drifting away as a cloud of atoms.

Electromagnetism (in the shape of the electromagnetic spectrum) is also responsible for the light that you see, the radio that you listen to, the microwaves that heat your food, and the ultraviolet rays that give you sunburn.

The electromagnetic force has a force-carrying particle we have already met in this book: the photon. Photons are massless particles that are basically tiny packets of energy that whizz along at the speed of light (which makes sense, since they *are* light).

GRAVITY

Along with electromagnetism, gravity is the only other force we consciously deal with in our daily lives. On the face of it at least, this most familiar of fundamental forces should be the easiest to get to grips with ... but appearances can be deceiving.

At its most basic, gravity is what makes things feel heavy and what makes stuff fall. Gravity is what gives mass its weight (mass and weight are not the same thing – an astronaut floating in space is weightless but he certainly isn't massless): without gravity to imbue matter with weight, there would be no stars, galaxies, or planets.

It is the only force that can be said to have an effect on a cosmological scale – the strong and weak forces can reach no further than a fraction of the span of an atom; and electromagnetism, despite being able to reach across the cosmos, pretty much cancels itself out by being equally balanced between positive and negative charges.

Gravity isn't constant over distance though, it follows the inverse square law – the gravitational attraction between two objects is proportional to their mass and inversely proportional to the square of the distance between them. In other words, if you take two massive objects and half the distance between them, their gravitational attraction increases fourfold.

When it comes to force carriers, gravity is the odd one out. Because it hasn't been successfully integrated into the Standard Model, physicists don't really know what its force carrier is, or if there even is one. So it doesn't feel left out, they've given it a hypothetical force carrier called a 'graviton', which (in theory) would be massless and light-speed quick.

We rejoin gravity in just a bit, but first we have to introduce our newest fundamental force.

THE HIGGS FIELD: THE OTHER FUNDAMENTAL FORCE

The discovery, in 2012, of a particle that appeared to be the Higgs boson (there is still a lot of work to do), has added another fundamental force to the list – the Higgs field.

The Higgs field was thought up by British physicist Peter Higgs in the 1960s as a solution to a rather crucial question: why do particles have mass?

There is no arguing with the success of the Standard Model – it has made countless successful predictions about how the counterintuitive quantum world of particles works – but it couldn't explain why the universe has mass.

According to the Standard Model, all the elementary particles that were born in the Big Bang shouldn't have been born with any mass at all.

As you can imagine, this was a crucial (and potentially embarrassing) omission because, without mass, there is no gravity and, without gravity, the roiling soup of particles spat out by the Big Bang would never have coalesced to form the stars and planets, or anything else. Since all evidence points towards the fact we do exist (leaving aside the existential musings of philosophers), this was a major problem.

The other problem was the discrepancy between the mass of different fundamental particles. The lightest elementary particle is the electron and the heaviest particle is the top quark, which is about 350,000 times more massive (about the difference between the Earth and the Sun). Logic would dictate that the top quark must be very much larger than the electron (like the Sun is much bigger than the Earth) – yet the particles are about the same size.

If a solution to these problems couldn't be found within the Standard Model, despite all its successes, physicists would have had to toss it in the bin and go back to the drawing board.

Peter Higgs' solution was the Higgs field – a sort of invisible dragnet which permeates the Universe. He proposed that as massless particles travel through the Higgs field they interact with it and, in doing so, acquire mass – the greater the interaction, the more mass they gain. So the reason the top quark is so much heavier than the electron is simply because it interacts more enthusiastically with the Higgs field.

WHAT IS MASS?

By definition, a particle without mass travels at the speed of light – it should really be called the speed of massless 'stuff', but, since the photon was the first massless particle to be discovered, the name 'speed of light' stuck.

Any particle that travels at less than the speed of light is therefore said to have mass. An object's mass is really just a description of how much effort it takes for something to change its speed – the greater its mass, the more effort it takes to accelerate (or decelerate) it.

HOW IS MASS DIFFERENT TO WEIGHT?

Mass is a measurement of the amount of matter an object is made of. Mass creates gravity and is a fixed quantity.

Weight is not fixed. It is a measurement of how much an object weighs within a gravitational field.

Take Professor Einstein for example:

On Earth he might weigh 70kg.

On the Moon his weight drops to just 12kg.

But, on a neutron star his weight will balloon to 7 billion tonnes.

His mass hasn't changed (he's still made of the same amount of matter), but the amount of gravity acting on his mass has – so his weight changes.

Just as the other forces have an associated force-carrier (or boson) particle, so does the Higgs field – arguably the first 'celebrity' particle: the Higgs boson.

To the relief of many physicists, and to the annoyance of some who were hoping for exciting new physics to replace the Standard Model, after 50 years (and

THE HIGGS FIELD

Massive attack: How the Higgs field gives particles mass

Particles gain mass by interacting with the Higgs field. The greater the interaction, the more mass the Higgs field imparts. This can explain how the top quark can be 350,000 times more massive than the electron – despite the two particles being the same size.

To explain this, we will dispose of our Lego metaphor and replace it with zombies (everybody loves zombies).

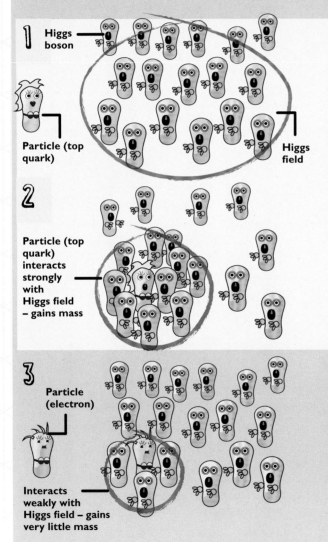

1 Higgs boson

Particle (top quark)

Higgs field

2

Particle (top quark) interacts strongly with Higgs field – gains mass

3

Particle (electron)

Interacts weakly with Higgs field – gains very little mass

1. Imagine the Higgs field is a room filled with particle zombies. The zombies represent the Higgs boson, and are spread out evenly across the room.

2. Our heroine, Miss T Quark comes running into the room. Zombies can't resist the allure of a sexy heroine so they gather round her (as zombies do). As they crowd around her, they slow her down (she loses momentum) and she gains mass.

3. Now a wooden doll, Miss Electron, is shoved into the room. Zombies want brains not wood, so they pretty much ignore the dummy. With very little to slow her down, the electron gains almost no mass.

So the more a particle interacts with the Higgs field, the more energy it loses and the more mass it gains.

tens of billions of dollars), the Higgs boson (or at least one sort of Higgs particle) was uncovered at the Large Hadron Collider in 2012.

But the Higgs isn't the whole story when it comes to mass. As we saw in the last chapter, electrons can be surrounded by a 'cloud' of virtual particles that 'pop' up from the quantum foam, whizz around for a while and then 'pop' off again.

Well, this is also true of all the other fundamental particles – the quark, for example, gets its mass from the Higgs field – but, when you add together the mass of the three quarks (and the gluons) that make up a proton or a neutron, the total falls well short of the total mass of the proton (or neutron). It's like building a spaceship from six Lego blocks that each weigh one gram and discovering the spaceship weighs 500 grams.

So, although the quark and gluon gets their mass from the Higgs field, most of the mass of the proton is made up of myriad virtual quark–antiquark pairs and gluons that flit in and out of existence.

In fact, about 95 per cent of a proton's mass (and a neutron's of course) comes from virtual particles, which means that 95 per cent of your body (and this book) is made up of stuff that only exists for a fraction of a fraction (of a fraction of a fraction ...) of a second.

So when someone says 'he's not all there', they might be closer to the truth than they realize.

GRAVITY OUTSIDE THE STANDARD MODEL (OR, GETTING HEAVY WITH EINSTEIN)

All the fundamental forces are very different but they act together to allow the Universe to exist and function. Physicists want to find a way to unify the forces within one theory that describes how they work and interact but so far they have failed.

It is one of the curious ironies of modern science that two of the most successful theories ever devised (and by theories we don't mean guesses) – quantum mechanics and General Relativity – seem to be quite incompatible. The strong force, the weak force, and electromagnetism have been unified within the Standard Model and it explains beautifully how these three forces work at the quantum level. But no one has found a way to bring Einstein's gravity into the quantum fold.

Einstein himself spent the last 30 years of his life on a fruitless quest to combine gravity with the other forces. He felt that there must be a single set of equations that could describe how nature works. In the 1920s, just four years after publishing his theory of gravity (General Relativity), Einstein set about trying to make the seemingly incompatible pieces of the forces jigsaw fit together within a single theoretical framework, which he called unified field theory.

Unfortunately, unbeknownst to him, the jigsaw was missing a few crucial pieces. Instead of a full set of fundamental forces, he only had electromagnetism and gravity to work with; and, rather than a full bag of fundamental particles, he only had the proton and the electron (the missing pieces wouldn't be discovered until the 1930s).

To further complicate the task, Einstein had rejected quantum theory (a theory that his work had helped found). He wasn't comfortable with the quantum weirdness, such as quantum uncertainty, that the theory threw up. As he famously quipped to physicist Niels Bohr: 'God does not play dice with the Universe'.

Suffice to say, Einstein failed to formulate his single unifying theory. Over the years many have tried and just as many have failed to bring

1 Leave your imaginary bowling ball to one side for a moment and roll an imaginary marble along the sheet. The marble will travel in a nice straight line.

2 Now pick up the bowling ball and place it on the sheet. The ball will distort the sheet and make a dent. When you roll the marble along the sheet now, it will still try to travel in a straight line but when it encounters the dent, the marble will curve around the bowling ball as if drawn towards it. In this case, your marble represents a photon, which, because it doesn't have any mass of its own and is moving at light speed, doesn't really fall into the dent. But, because the space it is travelling through has been distorted, its path is deviated slightly.

3 If you increase the mass of your marble and slow it down – so it represents a planet – it will fall further into the gravity well. If it has enough momentum, the marble will run around the edge of the depression – it will orbit the bowling ball star. If it has too much momentum, it will escape the well and be flung out into space. If doesn't have enough momentum, it will fall right into the well and collide with the star.

4 Now we will pull the bowling ball downwards to simulate a much more massive object. This time the gravity well is so deep that not even a photon will have enough momentum to escape – everything falls into the dent – just like a black hole.

This is gravity. All objects with mass distort space-time (even you and me) and all things in motion are subject to the influence of gravity – even light. The bowling ball isn't directly attracting the marble (as Newton imagined it) – instead it has changed the shape of the space through which the marble is travelling.

To think of it a slightly different way, you could say that gravity is really just 'falling with style'.

GETTING HEAVY WITH GRAVITY

A common analogy used to explain gravity as described by Einstein's theory of General Relativity is to imagine the fabric of the Universe (space-time) as being like a sheet of rubber and a massive object, such as a star, as being like a bowling ball.

Build your own space-time cliché

You will need the following imaginary objects:

• rubber sheet to represent space-time.

• bowling ball to represent a massive object (like a star).

• marble to represent a less massive object.

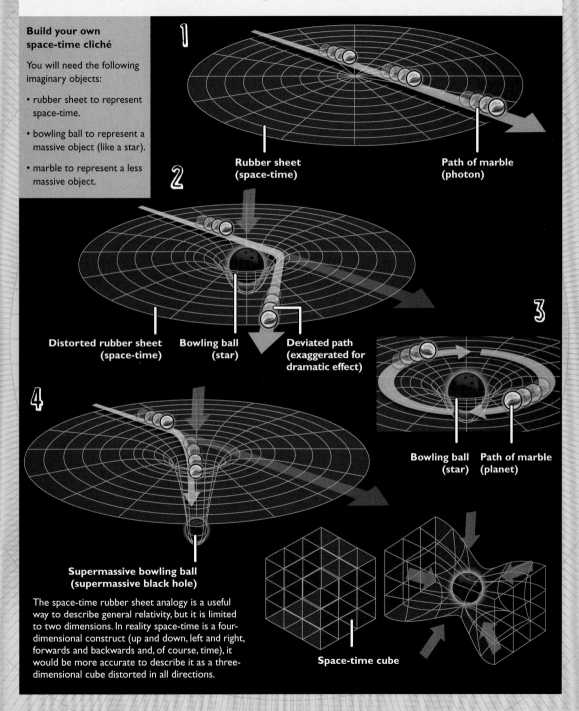

1

Rubber sheet (space-time)

Path of marble (photon)

2

Distorted rubber sheet (space-time)

Bowling ball (star)

Deviated path (exaggerated for dramatic effect)

3

Bowling ball (star)

Path of marble (planet)

4

Supermassive bowling ball (supermassive black hole)

The space-time rubber sheet analogy is a useful way to describe general relativity, but it is limited to two dimensions. In reality space-time is a four-dimensional construct (up and down, left and right, forwards and backwards and, of course, time), it would be more accurate to describe it as a three-dimensional cube distorted in all directions.

Space-time cube

MASSIVE DOESN'T MEAN HUGE

In day-to-day life and language, a massive object is one of enormous size but, in this book, it means something a little different. In physics, something doesn't have to be huge to be massive.

A neutron star, for example, can be about the same diameter as a large city but it is far more massive. In a neutron star, the particles are packed together so tightly that all the empty space has been squeezed out. If you could scoop out a teaspoon of neutron star material and take it to Earth, it would weigh about a billion tonnes.

the two together. Along the way, all sorts of increasingly weird and wonderful theories (and by theory we do mean guess, or, to be precise, hypothesis) have been cooked up: String Theory, M-theory and Super Symmetry to name but a few (we'll explore some of these later, too).

Although physicists are still trying to wrestle gravity into the barn of quantum theory, for our purposes, it is the kind of gravity described by Albert Einstein (and, to some extent, the version described by Isaac Newton) that we will use to build our Universe.

SO WHAT IS EINSTEIN'S GRAVITY?

For more than 200 years, Isaac Newton's Law of Gravity reigned supreme. According to Newton's theory, the force of gravity acts instantaneously – if the Earth were to become suddenly heavier, then every other object in the solar system would feel the change in the same instant.

But, in 1906, Einstein published his theory of Special Relativity, which showed that nothing can travel faster than the speed of light. The instantaneous nature of Newton's gravity was incompatible with special relativity's universal speed limit.

Then, in 1916, Einstein published his theory of General Relativity and overturned the Newtonian apple cart. For Newton, space was just the stage on which the laws of physics played their parts, but Einstein showed that space and time are also players on the stage. Mass causes space to curve and space causes mass to move.

According to Newton, gravity's effects are only felt by objects with mass, but Einstein showed that it can affect even massless objects like photons. In Newton's

THE TROUBLE WITH GRAVITY

Gravity has thus far resisted all attempts to make it fit into the Standard Model and play nice with the other fundamental forces.

Gravity has the power to bend space and time and tether entire worlds in orbit around stars, so you'd be forgiven for thinking that it must be a very powerful force indeed.

In fact, when compared to the other forces, it is astonishingly feeble. It is so weak you would need to increase its strength by a thousand billion, billion, billion, billion times to bring it in line with the strength of the other forces.

Gravity is so weak you can overpower it yourself

Just grab a nail and place it on a table.

Gravity is using all its strength to pull that nail as close as possible to the Earth's centre of mass.

Now take a small magnet and watch in awe as its electromagnetic force easily dismisses the gravitational force of an entire planet and lifts the nail.

It takes the mass of an entire star to overpower the electromagnetic force of a single atom. Only in the heart of a star is gravity strong enough to force atoms to fuse together.

EXTRA DIMENSIONS

To explain the mismatch between gravity and the other forces, physicists have suggested that there may be extra dimensions beyond the three that we are familiar with – up and down, left and right, forwards and backwards.

In the science fiction world, extra dimensions are depicted as alternate universes or parallel worlds (usual inhabited by the hero's evil alter ego – to be recognized by darkened eyes and a beard), but in physics an extra dimension is rather less dramatic.

For physicists, an extra dimension is just another direction in space on top of the three that we humans use to navigate the world. The extra dimensions are hidden from us because of the way we perceive the universe.

One theory, an offshoot of string theory, called M-theory (the 'M' stands for Membrane), predicts that there are up to ten dimensions and that the extra dimensions are hidden from us because they are curled up in really (really, really) small loops.

If that sounds bizarre, imagine an acrobat balancing on a tightrope. In essence, he is occupying a one dimensional world in which he can move only backwards and forwards.

Now, if we imagine a flea on the same tightrope. The flea can move backwards and forwards on the rope but he can also walk sideways and walk around the rope. The flea is living in a two dimensional world, but one of these dimensions is a tiny closed loop.

The acrobat can't detect the second dimension just as we can't detect dimensions beyond the three we move about in.

Also, just as we are trapped within our three-dimensional world, so is everything we use to measure

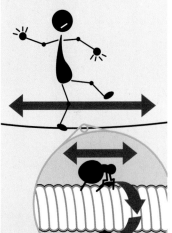

A tightrope walker can only move backwards and forwards on his rope. To him, he occupies a one-dimensional universe.

But a flea can move around the rope as well. To him it is a two-dimensional universe, but the extra dimension is too small for the tightrope walker to perceive.

the world around us – such as light and sound. With nothing interacting with these other dimensions, we have no way of detecting them.

So, although all the other fundamental forces are trapped in our three-dimensional world, gravity is thought to be free to travel through the extra dimensions.

As it spreads out through all the extra dimensions it becomes increasingly diluted – making its effect on our three-dimensional world much weaker.

TESTING EINSTEIN

When he laid out his theory of General Relativity in 1916, Einstein suggested a way to test gravity's light-bending effect. He said that, during a total solar eclipse, it might be possible to measure how the light from a distant star is bent by our local star, the Sun.

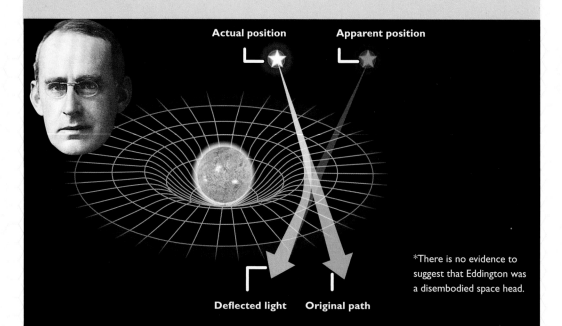

Actual position

Apparent position

Deflected light   **Original path**

*There is no evidence to suggest that Eddington was a disembodied space head.

A total eclipse on 29 May 1919 provided just such an opportunity. Led by British astrophysicist, Arthur Eddington (pictured above*), teams of astronomers were sent to the island of Príncipe, off the west coast of Africa, and to Brazil to observe the eclipse.

The position of a distant star (visible near the Sun) was photographed during the eclipse, when the Sun's glare was obscured, and then overlaid with a photograph of its position when the Sun was absent in that portion of the sky.

Both teams observed that the star's position appeared to shift – indicating that the Sun's gravity had indeed bent the path of the star's light.

Arthur Eddington's original image taken at Príncipe showing the apparent movement of the star.

gravitational world, light should be immune to the effects of gravity – but (as we will see later) light can be 'bent' by the gravitational pull of massive objects.

Rather than being an intrinsic force felt by objects with mass, Einstein showed that gravity is really a sort of by-product of the effect that massive objects have on the fabric of the Universe itself.

Every object with mass distorts the underlying fabric of the Universe, called space-time: the greater the object's mass, the greater the distortion is (see page 83). The effect is often likened to a bowling ball making an indentation in a sheet of rubber – any less massive object (like a marble) will roll towards the dent. It is the effect that these 'gravity wells' have on less massive objects (such as you and me) that we experience as the force of gravity.

How much an object is affected by gravity depends on its mass and the speed it is moving at. For a photon, moving at the speed of light, the distortion of space-time is really only enough to cause it to deviate slightly off course. But for the much more massive (and far more sluggish) Earth, the distortion is enough to prevent the planet from escaping the 'dent' and it is trapped in a circular (well, elliptical) orbit – it still wants to travel in a straight line but it is tethered to the Sun's gravity.

If the Sun were suddenly to vanish, the 'dent' would disappear and, a few minutes later (allowing for the journey time of the 'message' that the Sun has vanished, travelling at light speed), the Earth would whizz off in a straight line (or it would try ... there are lots of more massive planets making their own dents in the Solar System).

Gravity's strength obeys the inverse square law because that is the nature of the curvature of space-time around the object – the closer you get to the object's centre of gravity, the deeper the gravitational well you are falling down and the greater the force you feel (and, of course, the more massive the object, the greater the distortion, the deeper the well, and the greater the gravitational force).

The most gravitationally powerful object in the Universe, a black hole, creates a gravity well so deep that not even light has enough oomph to escape its clutches (but we are nowhere near to building a black hole yet).

OK, so we have our particles, our fundamental forces and we've had a Big Bang – it's time to bring some illumination to the Universe, dispel the darkness and build ourselves some stars.

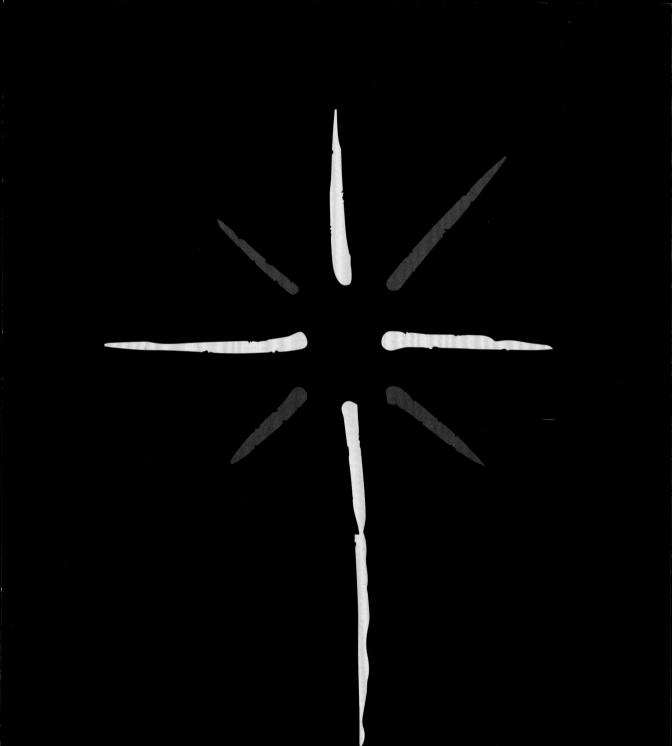

A STAR IS BORN

In which we gather atoms and marshal the fundamental forces to build the first complex structures, start the first nuclear fires, and turn the Universe away from the darkness.

S o here we are standing on the threshold of our new Universe. Let's pause a moment to take it all in.

OK, so there's not a lot to take in – the Universe is a dark, barren expanse of inky nothingness; a vast black desert of expanding space-time, bereft of life and features.

But, like a desert, its lack of features is an illusion. True, when viewed from an aircraft flying through the baking skies, a desert might appear to be a homogeneous expanse of nothingness, but look a little closer and all that changes.

As you move in to land your aircraft (watch out for sink holes), the plain beige ocean is transformed into a dynamic landscape of ever-shifting dunes. Now, leave your aircraft and get on your hands and knees. You will see a multitude of sand grains – billions of tiny silicate dancers, engaged in an intricate dance – touching and interacting in a dizzying waltz of near-infinite complexity.

In our Universe too, nothingness is never as absolute as it appears to be. Look a little closer. You will see all those atoms of hydrogen and helium we built in the Big Bang. Like the grains of the sand in our metaphorical featureless desert, our featureless Universe is populated by billions and billions (and billions and billions and billions) of atomic dancers just waiting for the music to start playing. As we discovered earlier, although it looks as if they are spread evenly across space-time from a distance (like the desert), the particles of matter are actually arranged a little more lumpily (like the sand dunes). Without that lumpiness, this would be the end of our little tale. The matter would continue cooling and the Universe would continue expanding – spreading it thinner and thinner into a future of eternal darkness.

Fortunately those lumps do exist and, as a result, so does our Universe.

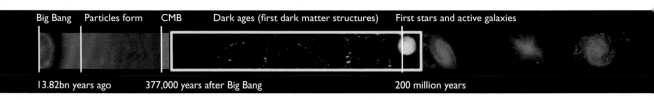

Big Bang   Particles form   CMB   Dark ages (first dark matter structures)   First stars and active galaxies

13.82bn years ago   377,000 years after Big Bang   200 million years

LET'S GET TOGETHER

ooooooooooooooooooooooooooooooooooooooooooooooooooooooo

So far we've used some of the fundamental forces to build the first atoms of simple elements, but this is where gravity really becomes the star of the show. We've seen how star-sized massive objects make gravitational 'dents' in the fabric of the Universe, but something doesn't have to be so big to bend space-time. Anything that possesses mass (even something as small as a single atom) makes its mark on space-time.

In today's Universe, there are so many large massive objects making large dents in space-time that an atom-sized dent is as irrelevant as a mote of dust next to the Himalayas. But our baby Universe contains nothing more massive than lithium atoms (which formed in small quantities alongside hydrogen, helium and deuterium), so you don't need a lot of mass to make a big impact.

Because our diffuse cloud of primordial hydrogen and helium is a bit 'lumpy', there are regions where the atoms are a little closer to each other and regions where they are further apart. Where they are closer, there is a little more mass making a slightly deeper dent – it's not a lot but even that tiny bit of extra gravity is enough to change our Universe dramatically.

Over millions of years, particles of atomic hydrogen and helium accumulate in the dents – forming increasingly dense clouds of gas with ever-increasing gravitational pull. More matter means more gravity and more gravity draws in more matter – it takes a while to get the ball rolling, but, once started, it becomes a runaway process (in cosmological terms at least – on yet-to-be-Earth, the dinosaurs will evolve, proliferate and go extinct in less time than it'll take for our first star to get going).

COSMIC TUG OF WAR

Even though we have some quite dense clouds of gas, which we might hesitantly start to call protogalaxies, we aren't ready to make a star just yet. The protogalaxies have more pressing priorities; they are fighting a tug of war with the expanding Universe.

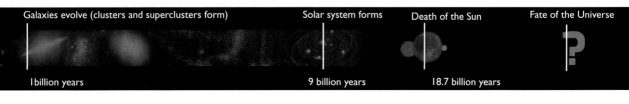

Galaxies evolve (clusters and superclusters form)　　Solar system forms　　Death of the Sun　　Fate of the Universe

1 billion years　　9 billion years　　18.7 billion years

About 200 million years has passed since the Big Bang and, although the speed at which it is doing so has slowed down, the Universe is still expanding. This means that the space between each of our hydrogen atoms is expanding also – the Universe is trying to tear our protogalaxy apart.

As it expands along with the Universe, the protogalaxy increases in volume, which is fine as long as it can keep sucking up gas from its surroundings, but, if it runs out of gas to feed on, it will be stretched thinner and thinner until it becomes a diffuse cloud once more.

But fear not, our protogalaxy has plenty to feed on and, when it does finally run out of food, it will have accumulated enough mass (between 100,000 and one million times the mass of the Sun) for its gravity to resist the expansion of the Universe. No longer bound to the expanding Universe, the protogalaxy can start to collapse under its own weight.

A DARK INTERLUDE

While our gas cloud is teetering on the brink of gravitational collapse and stellar ignition, we need to pause to consider a trifling flaw in the process I've just described: it shouldn't

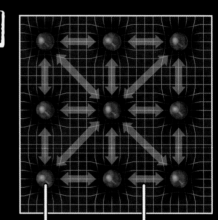

1. If this cloud had been spread perfectly evenly, gravity would have acted perfectly evenly on each particle within it. With each particle being pulled (and pulling) the same amount in every direction, they would have remained perfectly stationary.

Hydrogen atom   Gravitational attraction

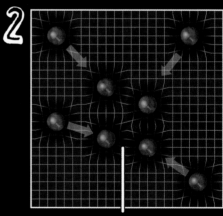

2. Luckily, as we saw in the cosmic microwave background, there were imperfections in the spread of matter and some areas were denser than others.

Region of increased density

3. Regions of higher density exert slightly more gravitational pull so particles in less dense regions are drawn towards dense regions.

Gravity well deepens and attraction increases

At the end of the recombination era, the Universe was filled with a diffuse cloud of gas – mostly made up of hydrogen.

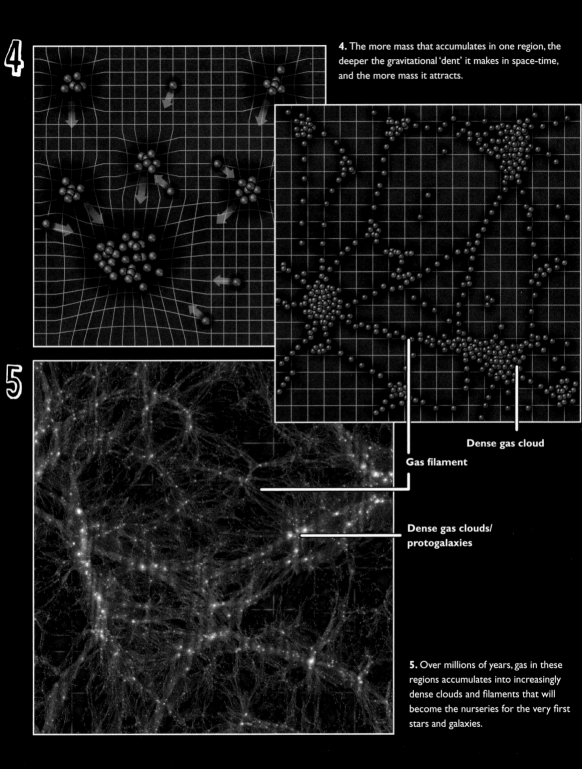

4. The more mass that accumulates in one region, the deeper the gravitational 'dent' it makes in space-time, and the more mass it attracts.

Dense gas cloud

Gas filament

Dense gas clouds/ protogalaxies

5. Over millions of years, gas in these regions accumulates into increasingly dense clouds and filaments that will become the nurseries for the very first stars and galaxies.

work. We've talked about how tiny ripples in the spread of matter created regions of increased gravitation, which prompted localized matter accumulation ... the trouble is, there wasn't enough matter in these fluctuations to get the process started.

If all we had to work with was the conventional matter that came from the Big Bang, at best it would have taken billions of years for it to gather enough gravitational oomph to start building stars, in which case you wouldn't be sitting here for another 4 billion years or so (if at all). At worst, there wouldn't have been enough matter to win the cosmic tug of war and expansion would have stretched the ripples and pulled them apart long before even the thinnest cloud of gas could have accumulated.

That you are sitting here reading this is evidence enough that this didn't happen, so something must be missing from our story – something that interacts

DARK MATTER HIGHWAYS

The rapid collapse of gas into the complex web of filaments and dense gas clouds couldn't have been achieved by the mass of normal matter alone – it needs the unseen mass of dark matter to provide the gravity we need.

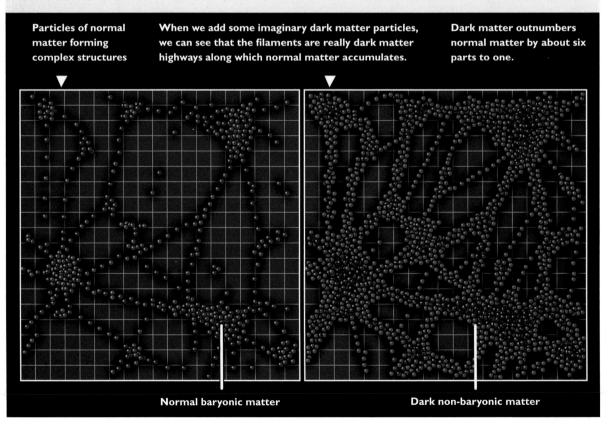

Particles of normal matter forming complex structures ▼

When we add some imaginary dark matter particles, we can see that the filaments are really dark matter highways along which normal matter accumulates. ▼

Dark matter outnumbers normal matter by about six parts to one.

Normal baryonic matter

Dark non-baryonic matter

DARK MATTER: HOW TO SEE SOMETHING YOU CAN'T SEE

So you've lost your dark matter bowling ball, bummer. Finding an object you can't detect directly is always a bit tricky. Fortunately, along with rubber sheet/space-time bending bowling balls, another side-effect of Einstein's relativity is something called gravitational lensing.

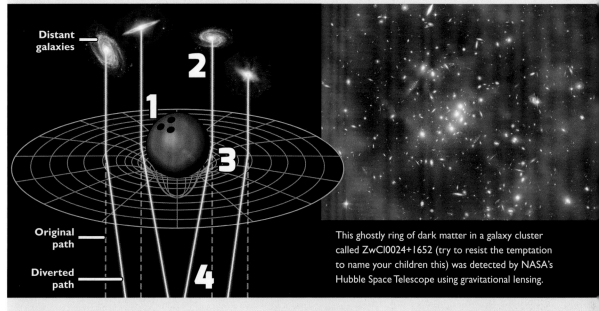

Distant galaxies

Original path

Diverted path

This ghostly ring of dark matter in a galaxy cluster called ZwCl0024+1652 (try to resist the temptation to name your children this) was detected by NASA's Hubble Space Telescope using gravitational lensing.

1. Here we have a dark matter bowling ball that, just like a bowling ball made of ordinary matter, bends space-time around it – creating a gravitational 'dent'.

2. Light from distant galaxies travels through space.

3. When all those photons encounter a strong gravitational field (what sort of matter it is created by is irrelevant), the light falls towards it and its path is diverted.

4. By looking for these light distortions, astronomers can build up a picture of dark matter distribution between us and the galaxies (and, of course, find your bowling ball).

with gravity but is otherwise invisible to us. That 'something' is the ever-mysterious substance known as dark matter.

The name 'dark matter' reads like a substance plucked from the fevered imagination of a low-grade science fiction writer, but, far from being a USS-*Enterprise*-endangering fictional construct, it is very real. Despite its name, it isn't matter's sinister alter ego (that, arguably would be antimatter, but it'd probably be unfair to call that sinister, too): it's called *dark* matter because we can't see it.

More precisely, dark matter is a form of matter that doesn't interact with the 'normal', or baryonic, matter that makes up the Universe we *can* see. It is immune to the charms of the electromagnetic force, which is why we can't detect it (our eyes, and the telescopes they peer through, rely on the electromagnetic spectrum). We *do* know it exists though because it does interact with gravity, which means we can detect its presence and influence by its gravitational effect on the normal baryonic matter that we can see.

Its existence was first suspected as far back as 1933, by a Swiss-American astrophysicist, Fritz Zwicky, who had been studying galaxy clusters (groups of galaxies bound together by gravity). He observed the motions of the galaxies within the cluster and applied Newton's laws to estimate its gravitational mass. But when he came to estimate the amount of visible mass within one particular cluster (by measuring the light emitted by the stars within it, extrapolating their mass and adding it all together) his figures fell drastically short of his first estimate: the visible mass accounted for only a fraction of the cluster's gravitational mass.

Furthermore, there wasn't enough visible mass to generate the gravity needed to hold the cluster together (the galaxies should have been flying apart, but they weren't). He concluded that there must be something invisible and undetectable making up all that missing mass and keeping the galaxy gravitationally bound together – and that something was dark matter.

To tell the full story of dark matter's discovery and the subsequent hunt for its constituent particles would require a book of its own, so, for now, we'll have to satisfy ourselves with the knowledge that dark matter makes up the vast majority of the mass of the Universe. We don't know what it is yet, but it outweighs normal matter (as described by the Standard Model) by about six to one.

It might seem a tad frivolous to just skip over something that makes up so much of the cosmos, but it is really only the gravitational influence that dark matter has on our Universe that matters (no pun intended) in this book, and to explore it any further would invite questions too deep to answer here (and, besides, it seems to have gone to a great deal of effort to stay out of the limelight and we wouldn't want to intrude on its privacy).

Anyway, all the extra gravitational mass we get from dark matter is just what we need to get all that gas collapsing into nice star-making clouds before the expanding Universe can tear it all apart.

So, now that we throw about six handfuls of dark matter for every single handful of normal matter into the mix, we have a sort-of dark matter scaffold, around which normal matter can gather.

FEELING THE SQUEEZE

OK, so let's go back to our protogalaxy – a vast cloud of ordinary matter (hydrogen, helium and a touch of lithium) and dark matter all mixed up, collapsing and compressing under the ever-increasing weight of its own gravity. As the cloud collapses and gas 'falls' into the ever-deepening gravity well, the atoms of hydrogen (for now, we'll just assume the presence of helium and lithium) gain energy and accelerate. Packed full of kinetic energy, they start smashing into each other, releasing their stored-up energy as thermal energy (heat) – the hydrogen clouds start getting hotter and hotter.

But then something strange happens: when the clouds get to about 800°C, the gas suddenly starts to cool down again. As the hydrogen atoms are forced closer and closer together, they begin to interact and their electron clouds become intertwined. Their nuclei don't fuse (that's going to take a lot more heat and pressure), but their individual electrons team up and start orbiting the multiple nuclei – creating hydrogen molecules.

One particular type of molecule, triatomic hydrogen, consists of three (tri) hydrogen nuclei (atomic) but is orbited by just two electrons. The imbalance between the positive

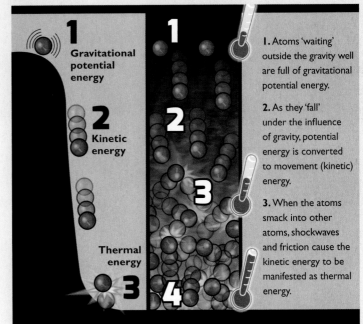

HOW GRAVITY CREATES HEAT

Anything with the potential to fall under the influence of gravity possesses stored 'potential' energy that can be transformed into heat.

1 Gravitational potential energy

2 Kinetic energy

3 Thermal energy

The closer the atom gets to the centre of gravity, the more potential energy is converted to kinetic energy.

1. Atoms 'waiting' outside the gravity well are full of gravitational potential energy.

2. As they 'fall' under the influence of gravity, potential energy is converted to movement (kinetic) energy.

3. When the atoms smack into other atoms, shockwaves and friction cause the kinetic energy to be manifested as thermal energy.

4. As pressure increases, more atoms are flying around with more energy, and colliding more often. The temperature increases rapidly.

SOWING THE SEEDS OF THE STARS

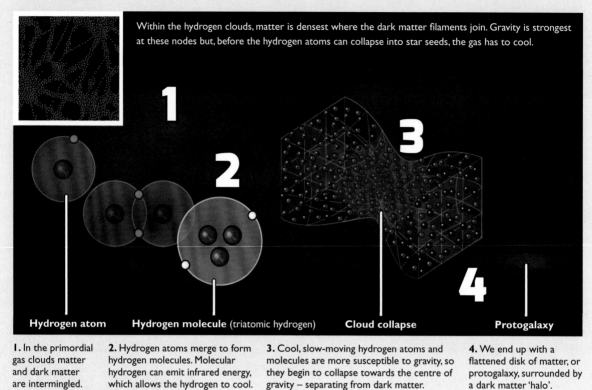

Within the hydrogen clouds, matter is densest where the dark matter filaments join. Gravity is strongest at these nodes but, before the hydrogen atoms can collapse into star seeds, the gas has to cool.

1

2

3

4

Hydrogen atom **Hydrogen molecule** (triatomic hydrogen) **Cloud collapse** **Protogalaxy**

1. In the primordial gas clouds matter and dark matter are intermingled.

2. Hydrogen atoms merge to form hydrogen molecules. Molecular hydrogen can emit infrared energy, which allows the hydrogen to cool.

3. Cool, slow-moving hydrogen atoms and molecules are more susceptible to gravity, so they begin to collapse towards the centre of gravity – separating from dark matter.

4. We end up with a flattened disk of matter, or protogalaxy, surrounded by a dark matter 'halo'.

and negative charges gives triatomic hydrogen molecules an overall positive charge and makes them very excitable. As they are struck by all the non-molecular hydrogen rushing around within the cloud, the charged molecules get all wound up (like a toddler who's found Red Bull in his tippy mug instead of milk), start vibrating and, instead of holding on to all that energy, they emit it as infrared photons – cooling the cloud down.

This sounds like a disastrous development (after all, if there's one thing a star needs, it is heat) but it is, as it happens, a crucial stage. A hot gas is subject to thermal pressure that pushes against the inward-pulling force of gravity and our cloud doesn't have enough gravitational clout to overrule this outward pressure.

By throwing photons (and therefore heat energy) out of the cloud, the molecular hydrogen causes the gas to cool and separate from the dark matter, which, because it doesn't interact with the electromagnetic force and isn't interested in photons, can't

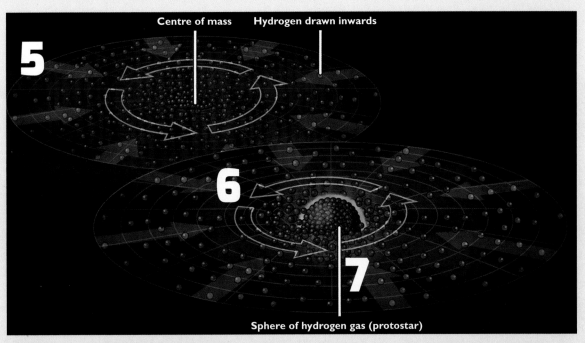

Centre of mass **Hydrogen drawn inwards**

5

6

7

Sphere of hydrogen gas (protostar)

5. Within the protogalaxy, regions of slightly cooler, denser gas collapse.

6. As the gas 'falls' into the gravity well, it begins to slowly rotate around the centre of gravity (like water falling down a plug hole) and flatten out into a disk.

7. At the heart of the disk, a protostar develops. As more matter piles inwards, the pressure increases and the temperature at its core rockets.

As the core gets hotter, the molecules break down and the atoms are stripped to their constituent protons and electrons. When this plasma reaches 15 million°C, nuclear fusion begins.

get hotter or colder. The more the atoms of normal matter cool and slow, the more they feel the tug of gravity and they start to sink to the centre of the cloud – leaving the protogalaxy as a flattened disk of slowly rotating matter encased in a blanket of dark matter (even today, galaxies are surrounded by a 'halo' of dark matter).

IGNITION!

Now that our hydrogen atoms are all nice and cool and (more importantly) gathered in one place, things can really heat up again. At the mercy of gravity like never before, all those slow-moving atoms form localized pockets of denser gas that, in turn, begin to collapse into denser clumps, called protostellar clouds.

Because gravity wants to pull everything towards a single centre of mass, the heart of the protostellar cloud becomes increasingly dense and hot – forming a sort of star seed, or protostar, whose ever-increasing gravitational pull siphons up the gases in the cloud.

As the gases fall into the gravity well of the protostar, the cloud begins to flatten out and (like water draining into a plug hole) it starts to slowly rotate. All the while, the protostar is gaining mass, collapsing and getting hotter and hotter and, since the centre of mass is the densest part, its core is the hottest part of all.

When the core approaches 2,000°C, all those hydrogen molecules are torn apart and returned to their constituent hydrogen atoms. Next, the hydrogen atoms themselves are stripped of their electrons (known as ionization) and the gas is transformed into a plasma of fast-moving protons and electrons (mirroring the conditions shortly after the Big Bang from which these atoms were initially formed).

Fuelled by the energy from gravitational collapse, the protostar becomes intensely hot and, at about 15 million degrees Celsius, the protons in its core have enough energy for nuclear fusion to begin.

Now, we can't just stick those protons together – even at 15 million degrees Celsius, the protons don't have enough energy to force their way past the repulsive barrier of their mutual electromagnetic repulsion (called the Coulomb barrier). Luckily, we can call on some of that quantum weirdness to help us out.

If you remember, at a quantum level, particles can be said to exist as a cloud of probability. As a smear of non-committal location, the protons' wave functions can overlap – allowing them to bypass the Coulomb barrier in a process called quantum tunnelling, where the strong nuclear force can take over.

Once two protons are joined in quantum matrimony, two things can happen: a proton pair is created, called a diproton; or one of the protons decays into a neutron via the weak force.

By far the most common result of the union is the creation of a diproton, but, sadly, diprotons are the subatomic equivalent of Richard Burton and Elizabeth Taylor – their union is very unstable and they quickly separate. But, about every billion years or so, one of the protons will decay into a neutron, making a stable deuterium (heavy hydrogen) nucleus. This might sound like an awfully long time to wait, but remember, there are trillions and trillions of protons whizzing around, so it won't take too long for one of the protons to make the change.

THE PROTONS AND THE QUANTUM TUNNEL

(a tale of quantum mechanics)

Once upon a time, there were two hydrogen atoms who desperately wanted to be joined in nuclear matrimony.

So they travelled to the church of Reverend Strong Force. But there was a problem.

The evil Count Coulomb had vowed to keep all protons apart and had erected an impassable electromagnetic force field to repel them.

Try as they might, the two protons just couldn't breach the walls. They built ladders to scale them, but no matter how fast they ran up them, they always ran out of energy before they reached the top.

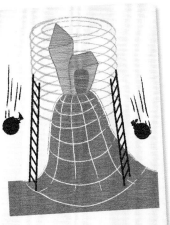

Then one day, Wizard Heisenberg arrived and said that he could help them. At first the pair were sceptical (mostly because the wizard didn't seem to have the least idea of who he was, or where he was going), but he gave them a magic spell that would allow them to magically tunnel through the walls.

'I call it quantum tunnelling,' he announced (although he didn't seem entirely certain).

So they ran up their ladders once more and used the uncertain wizard's spell.

The spell revealed the dual wave/particle nature that every particle possesses in their hearts and they were transformed into a cloud of probability.

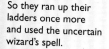

As wavefunctions, the protons could exist at any location in the magic quantum cloud. So, travelling as waves, the two protons approached the electromagnetic barrier. Because they didn't exist at any particular point in the wave, parts of it could overlap with the barrier.

The uncertain wizard then said a magic word, which forced their wave-like forms to collapse and their particle forms to appear where they had overlapped the barrier.

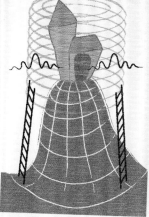

Together at last, they were joined by the Reverend Strong Force and, where once there had been two separate protons, there appeared a single deuterium nucleus.

The end
(or is it?)

Deuterium nuclei are quite keen to attach with another proton (who said a successful marriage depends on monogamy?). It'll only take a second or so for another proton to collide with the deuterium nucleus, and the strong force will have no trouble holding them all together to make a helium-3 (light helium) nucleus.

In another half million years (give or take a millennia or two) this nucleus will collide with another helium-3 to form the more familiar helium nucleus of two protons and two neutrons. This new union releases two high-energy gamma ray photons, two protons – starting the process all over again (hence the name, 'proton-proton chain') – and a huge amount of energy. But where does this energy come from?

When all those protons and neutrons are fused together to make heavier nuclei, they lose some of their mass, and Einstein's famous $E = mc^2$ equation

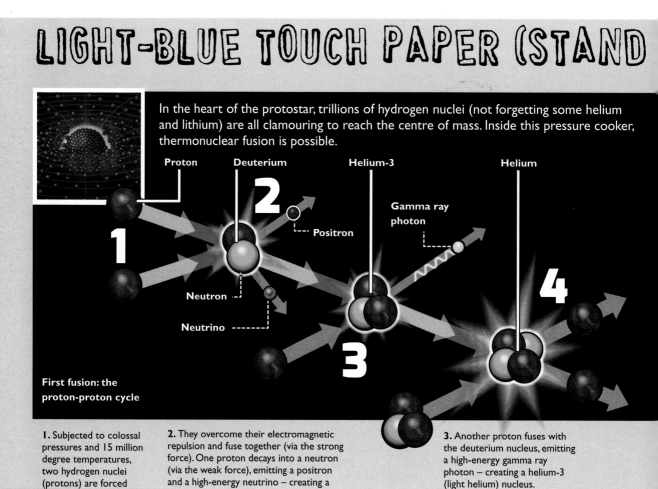

LIGHT-BLUE TOUCH PAPER (STAND

In the heart of the protostar, trillions of hydrogen nuclei (not forgetting some helium and lithium) are all clamouring to reach the centre of mass. Inside this pressure cooker, thermonuclear fusion is possible.

First fusion: the proton-proton cycle

1. Subjected to colossal pressures and 15 million degree temperatures, two hydrogen nuclei (protons) are forced together.

2. They overcome their electromagnetic repulsion and fuse together (via the strong force). One proton decays into a neutron (via the weak force), emitting a positron and a high-energy neutrino – creating a deuterium, or heavy hydrogen, nucleus.

3. Another proton fuses with the deuterium nucleus, emitting a high-energy gamma ray photon – creating a helium-3 (light helium) nucleus.

(Energy = mass × the speed of light × the speed of light) tells us that mass lost is converted to energy. Only a tiny amount of mass is lost in each reaction – about 0.7 per cent (and 0.7 per cent of an atom is very small indeed), but so many reactions take place within the core of a star that, in a Sun-size star, 600 million tonnes of hydrogen are converted into helium every second. So every second the Sun 'loses' 4.3 million tonnes of mass, which converts into a lot of energy.

Energy lost as gamma rays will interact with electrons and protons to create heat and all those positrons will interact with electrons and annihilate to release even more energy. Combine all that with the kinetic energy carried by the protons thrown out from the reactions, and you have created a lot of heat. The neutrinos created as another by-product carry a lot of energy too, but, because they hardly

WELL BACK)

4. Finally, two helium-3 nuclei fuse to create a helium (helium-4) nucleus. Two protons are ejected along with lots of energy.

5. The first star flares into life. For the first time since the Big Bang, new light and energy is coursing through the cosmos.

6. These early, massive stars burn furiously as hot blue giants, but their very fury means their days are already numbered.

react with their surroundings, they shoot off into space, unimpeded by the gas that surrounds them.

With the ignition of nuclear fusion, what was once a diffuse cloud of slowly cooling gas has been transformed into a blinding furnace of roaring, incandescent fury. The atoms that became a cloud, that became a protostar, have completed the first part of their journey and become a star. For the first time since the Big Bang, the Universe is illuminated: its glory has only just begun.

BIG BABIES

Just how big the very first stars were is a matter of some debate. Until recently, it was thought that they must have been truly massive – stellar behemoths that weighed the same as hundreds (and maybe thousands) of Suns. This was based on the idea that those first star-forming clumps were much warmer than the clouds of molecular hydrogen in which stars form today. In present-day star-forming regions, the clouds are cooled by dust grains, and molecules containing heavy elements. But in our infant Universe, those heavy elements just didn't exist.

HEAT: HOLDING UP THE FORT

Nuclear fusion creates a lot of heat energy so you'd think that, once this is combined with all the heat being generated from gravitational collapse, the core would get hotter and hotter and hotter (after all, on its own, that was enough to heat the core up to 15 million degrees Celsius). Fortunately for our new-born star, that isn't the case. If it was, the star would get so hot that it would tear itself apart in a colossal star-destroying explosion before you could say 'blimey that's hot'.

Ironically, the energy released by fusion actually has a cooling effect by 'pushing' against gravity and preventing the star from collapsing under its own weight. It's a finely-tuned balancing act that can support a star for billions of years, but, when it breaks down (and it always does) the star is doomed.

Because the clouds have to cool before they can collapse, it was thought that those hot early clouds would have needed a lot more mass (about a thousand times more) to overcome their thermal resistance and get the gravitational ball rolling. Recently, this assumption has been questioned. Astronomers have failed to find evidence of these stellar giants and computer simulations have showed that, although it did require more mass to get the ball rolling, the earliest protostars became so hot (about nine times hotter than our Sun) that their thermal pressure quite quickly overcame gravity – blowing gas into space and preventing the star from becoming overweight.

Despite their recent downsizing, the first stars were still massive beasts that weighed in at many tens of times the mass of the Sun. Our new star is still a very big baby (oh, and it's blue ... like the hot part of a gas flame, the hotter the star, the bluer it shines).

CONGRATULATIONS IT'S TWINS!

Another long-standing assumption brought into question in recent years is the idea that the first stars were all cosmic loners and, again, the revelation comes from computer simulations.

The simulations, which involve creating virtual protostellar clouds and subjecting them to simulated versions of the conditions thought to exist at the time, seem to show that the majority of early stars likely formed not in isolation, but in tight systems of multiple stars.

Research suggests that, as a protostar pulls in matter and creates a stellar disk, the disk becomes so massive that it becomes gravitationally unstable and fragments. These fragments then go on to collapse to create more protostars. These 'companion' stars remain gravitationally bound to the first protostar, which remains the most massive, forming double (binary) or triple (trinary) star systems that orbit closely together around their shared centre of mass.

The complexity of the computational task means that even supercomputers can only take the simulations as far as the first 10,000 years of stellar development, so the research does have its limitations, but it could explain why 80 per cent of the stars that live in the Universe today belong to multiple star systems. Our lonely Sun is a bit of a cosmic oddball.

AND THEN THERE WAS LIGHT (AGAIN)

○○○

If you think back a couple of million years (or a few pages), you might recall that I announced (with some fanfare) that 'at last light was free to travel the cosmos'. Well, I have a confession to make: light wasn't as free as I may have implied.

All the neutral hydrogen made by electron capture during the (misleadingly named) recombination era might have been transparent to low-energy photons (infrared, radio, microwaves, etc) but it was quite opaque to higher-energy photons like visible light and ultraviolet radiation. Light photons in these frequencies are easily absorbed by neutral hydrogen atoms: although the first nuclear beacons have been lit, the stars are hidden behind a smoke screen of opaque hydrogen.

But stars have a knack for getting themselves noticed and nothing as trifling as gas is going to stop them. Because our star is burning so very fiercely, it is throwing huge amounts of high-energy radiation (mostly ultraviolet) into the space around it and this does a very good job of heating up the gas around the star. As we've already discovered on several occasions, when neutral gases are heated they lose their grip on the electrons that orbit their nuclei and become ionized.

Now I'm going to seem to contradict myself again, but bear with me ... Cast your mind back once more to the recombination era and you will recall that I said it was only when the free electrons flying around in the ionized gases had been captured by atomic nuclei that the Universe first became transparent. Well, this time the reverse is true, and when the first stars reionize the gases that surround them, this allows photons of visible light to pass through them. But the difference is: the gases were much, much denser back then – there were so many protons and electrons flying around that the gases formed a dense, impenetrable plasma. Crucially, reionization set off the decoupling era (unlike recombination, this era deserves its prefix) – when the matter and radiation that, until then, had been tightly coupled (bound in interactions) became free to do their own thing. In the thin clouds that remain around our star, reionization actually removes the last barrier to the movement of light in the cosmos.

The first stars do little more than blow localized bubbles into the fog, but, as star ignition accelerates over the following millennia, these bubbles will join together until vast regions of space become transparent. Universe-wide reionization will take about a billion years (and it'll take the formation of something rather more spectacular than a simple star to get the job done), but the formation of our first star heralds the beginning of the end for the Universe's dark ages.

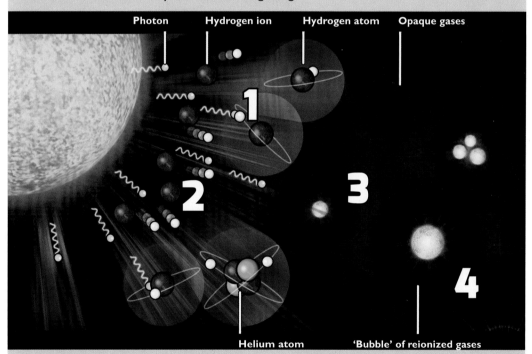

THE END OF THE DARK AGES

The era of reionization began with the formation of the stars, about 200 million years after the Big Bang, and ended when the majority of the opaque neutral gases had been ionized, about one billion years after the Big Bang.

Photon   Hydrogen ion   Hydrogen atom   Opaque gases

Helium atom   'Bubble' of reionized gases

1. Every atom has particular frequency of radiation that it likes to absorb. For neutral hydrogen, that just happens to be light in the visible and ultraviolet part of the spectrum. These photons are absorbed, which blocks the passage of light.

2. The first stars heated up the gas around them – causing hydrogen atoms to lose hold of their electrons. The ionized hydrogen is transparent to visible light and infrared photons.

3. Each new star created a bubble within the gas (known as the intergalactic medium) of ionized hydrogen around it.

4. As more bubbles formed, they grew and came together, burning away the fog of the dark ages.

THE LIFE AND DEATH OF A STAR

In which we put our new-found hydrogen fusing skills to use and employ the stars as cosmic pressure cookers to bake up some heavy elements (and we kill a few stars and make a black hole).

Who could fail to be impressed with our achievement so far? In just a few short eons we have taken an infinitesimally small speck of energy and transformed it into a fledgling Universe, filled with baby blue stars and fuzzy little juvenile galaxies. And all we've had to do to achieve this is summon matter and energy from the quantum foam; lay an ever-expanding carpet of four-dimensional space-time; craft matter from energy and squeeze energy from matter. It's a wonder more people don't do it.

But let's not get carried away – we've done very well, but before we start declaring our omnipotence and dictating books for superstitious entities to build spurious frameworks for creator-worship around, we need to add a little complexity to our creation.

Despite our achievements to date, we still have the same quantities of chemical elements that we had when they coalesced from the roiling inferno of the Big Bang – about 75 per cent hydrogen and 25 per cent helium, with trace amounts of deuterium, helium-3 and lithium.

As I am sure you're aware, as useful as these elements are, most of what we consider to be essential Universe furniture (planets, moons and life itself) is built from elements like carbon, oxygen, nitrogen and iron. Somehow we need to transform humble atoms of hydrogen and helium into atoms of heavy elements.

Fortunately, we have just built the ideal crucible for cooking up some alchemical magic – the stars. Billions of years from now, the inhabitants of yet-to-be-Earth will spend centuries locked away in blackened rooms, making notes in secret codes and inhaling all kinds of volatile lung-wrecking compounds; all in a futile attempt to transform base elements into gold. Little did they know that all they really needed was a many-trillion-tonne ball of thermonuclear plasma to get the job done.

HOW MANY MILLENNIA TO THE GALLON?

The burning of hydrogen is the life blood of every star in the Universe – it is the easiest to fuse into heavier elements and, because the reactions liberate the most energy, it gives the best 'bang for the buck'.

The very first stars are thought to have been extremely massive (many tens of times the mass of our Sun) and, of course, more mass means more hydrogen. Now, you'd be forgiven for thinking that, with all that extra hydrogen knocking around, a very massive star would last a very long time, but, in fact, the very opposite is true: more mass actually equals a shorter life. This might sound a little counterintuitive, but it does make sense when you think about it.

Fusion reactions are made possible by the huge gravitational pressure created as the rest of the star's mass squeezed down on its core. The greater the mass of the star, the more gravitational pressure is exerted on the core and more furiously it has to burn hydrogen to create the heat needed to push outwards against the crushing force of gravity.

Massive stars might begin life with lots of extra fuel, but they burn through it absurdly quickly. A star with perhaps 50 times the mass of the Sun will exhaust its supply in as little as a million years, yet our local star, which has already been burning for 4.6 *billion* years, is only now approaching middle age. You could compare it to the difference between a tiny city car and an American muscle car – the city car might only carry a few gallons of fuel compared with the muscle car's 20 gallon tank, but, because it uses it more efficiently, the smaller car will still be running long after the muscle car has spluttered to a petroleum-starved stop.

So, our first massive star has a rather depressingly short life expectancy, but its thermonuclear fury makes it a powerful alchemy machine, custom-built to transform hydrogen into all sorts of heavy, multi-proton goodies.

THE LIFE CYCLES OF THE STARS

A star's life cycle is determined by its mass and how hot it burns – supermassive stars can burn through their nuclear fuel in as little as a few tens of thousands of years, yet stars just a fraction of their size will burn for many times the current age of the Universe.

Brown dwarfs are often described as failed stars. They formed from protostars that didn't have enough mass to ignite full-blooded hydrogen fusion. All they can look forward to is a slow death as they haemorrhage heat into the vacuum of space and fade from view. More like huge gas giant planets than stars, I like to think of them as over-achieving planets.

Red dwarfs are tiny but have enough mass to make hydrogen fusion possible. But they burn at such low temperatures that they'll still be dimly shining away when Universe is many times its current age. The huge majority of all the stars in the Universe are red dwarfs – accounting for about 75 per cent of all stars.

Sun-like stars (or yellow dwarfs) have enough mass to ignite hydrogen and helium fusion in their cores. When they run out of helium they expand to become red giants, shed their outer gas layers as planetary nebulae and leave behind a white dwarf star. Over hundreds of billions of years (if the Universe lives long enough), they will slowly cool to become black dwarfs.

Supergiants and hypergiants are the morbidly obese members of the stellar community. With masses ranging from tens of Sun-sized stars to many hundreds, they burn through their enormous fuel reserves in as little as a few tens of thousands of years.

Creators of all the heavy elements in the Universe, once iron has been created in their cores, they explode as supernovae.

Less obese stars will end their lives as neutron stars or pulsars, but the morbidly obese become crushed into oblivion by their own colossal bulk and become black holes.

Sizes on this graphic are not even slightly to scale. A supergiant with 20 times the mass of the Sun, for example, would be about 75 times larger than the Sun.

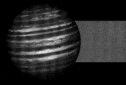

Brown dwarf
Mass: 0.08x (solar masses)
Temperature (surface): 1,000°C
Life expectancy
(main sequence): n/a

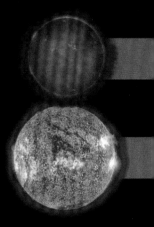

Red dwarf
Mass: 0.2x
Temperature: 3,000°C
Life expectancy:
10 trillion years

Sun-like star
Mass: 1x
Temperature: 5,000°C
Life expectancy:
10 billion years

Supergiant
Mass: 20x
Temperature: 12,000°C
Life expectancy:
5 million years

Hypergiant
Mass: 100x
Temperature: 40,000°C
Life expectancy:
1 million years

Stars in the prime of their lives are known as main sequence stars. During this period they are burning fuel at just the right rate to remain stable. Usually the end of main sequence life is marked by the star cooling and expanding to become a red giant.

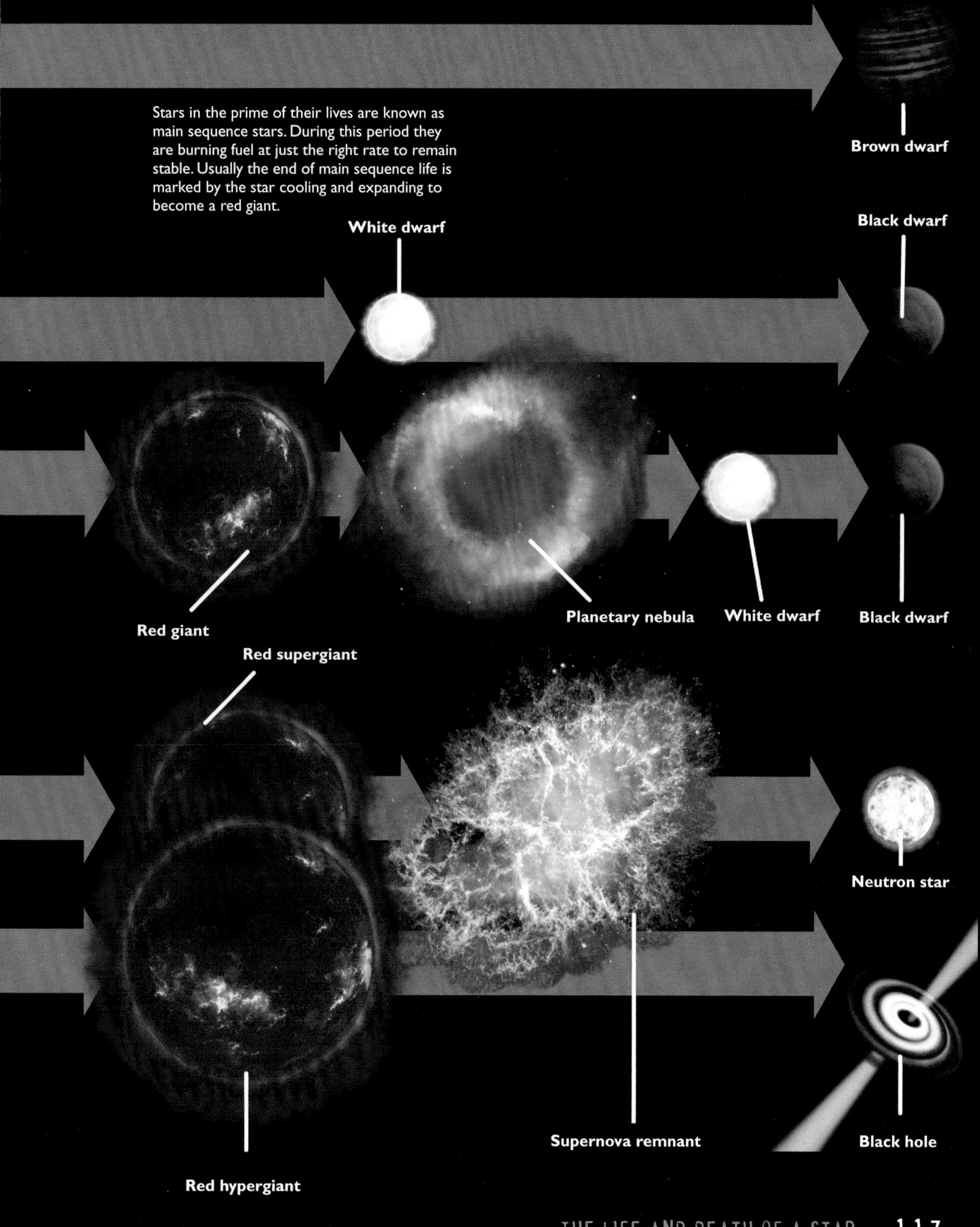

Brown dwarf

White dwarf

Black dwarf

Red giant

Planetary nebula

White dwarf

Black dwarf

Red supergiant

Neutron star

Supernova remnant

Black hole

Red hypergiant

HEAVY METAL GODS

●●●

The process by which light atoms are fused to create heavier elements is called nucleosynthesis. We've already seen how hydrogen is fused via the proton-proton cycle to create helium, but what happens when a star exhausts its hydrogen stockpiles?

How quickly a star runs out of hydrogen depends on its mass, but, whether it burns through it in a few hundred millennia or ekes it out for billions of years, every star runs out of hydrogen eventually.

When a star exhausts its hydrogen, fusion reactions shut down inside the core. Without the outward pressure created by heat pumping from the star's centre, the delicate balance of internal thermal pressure and gravity is broken and, with nothing to hold it back, all the mass that has been clamouring to fall inwards does just that and the core collapses under the weight of its own gravity.

As it collapses, pressure increases in the core and it starts to heat up once more. It passes the 15 million degrees Celsius threshold for hydrogen fusion and, because there's no hydrogen left, it keeps on going. The core gets denser and denser and hotter and hotter until it reaches the next magic temperature: 100 million degrees Celsius.

At this temperature, the stage is set for the next phase of our star's life: helium fusion. In this phase of fusion reactions, helium-4 nuclei are 'stuck' together to create heavier elements whose nuclei contain protons and neutrons in multiples of four and we end up with elements like carbon-12 and oxygen-16 (see pages 116–17). This might sound as simple as sticking together a few four-stud Lego bricks to get a stack with 12 or 16 studs, but, like the hydrogen fusion process that preceded it, helium fusion is much more complicated than just squeezing together a few helium nuclei.

If you fuse two helium-4 nuclei together, you get a beryllium-8 nucleus. But you can't stick two of these together to get an oxygen-16 nucleus because beryllium is so unstable that it falls apart long before you can bring two of them together. In the first half of the 20th century, this caused quite a headache for astrophysicists – they knew heavier elements existed (duh!) but couldn't figure out how that was possible.

Then, in the 1950s, Fred Hoyle (he of Big Bang coining fame) figured out that a stable carbon-12 nucleus could be made by sticking three helium-4 nuclei together all at once (called the triple-alpha process because helium nuclei are also known as alpha particles). We don't have the space to construct a detailed view of the triple-alpha process, but I mention it because it was a profound and hugely important piece of deductive reasoning. Even though the prevailing wisdom proclaimed the process to be impossible, Hoyle persevered because he knew carbon-12 existed, therefore there *must* be a process that made it possible. Or as he said: 'Since we are surrounded by carbon in the natural world and we ourselves are carbon-based life, the stars must have made a highly effective way of making it, and I am going to look for it.'

His discovery revealed a peculiar (and very important) coincidence. It turns out that carbon-12 has just the right combination of quantum attributes to allow it to bypass the unstable beryllium-8 stage: if just one of these variables had been out of kilter, stellar fusion would have ended with the formation of beryllium and you and I wouldn't exist. Once again we find ourselves marvelling at our Universe's 'against the odds' existence.

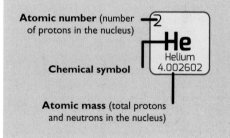
Eventually, of course, the star exhausts its helium reserves and fusion shuts down once more. Leaving behind a thin shell of unburned helium (there's also a thin layer of unburned hydrogen around that) the core shrinks yet again, making it hotter and hotter until it reaches the 600 million degrees Celsius temperature required for carbon fusion. Carbon fusion is a much more straightforward process than helium fusion. In this reaction, two carbon-12 nuclei fuse to create the nuclei of heavier elements such as magnesium-24.

The heat from the latest round of fusion stabilizes the core and halts its collapse again. But, inevitably, the carbon supply becomes exhausted and the process of collapse, unburned fuel shell deposition, heating and fusion ignition starts again. Each fusion phase requires higher temperatures and creates successively heavier elements all the way up to iron-56, but this is where the proverbial 'buck' is stopped in its tracks.

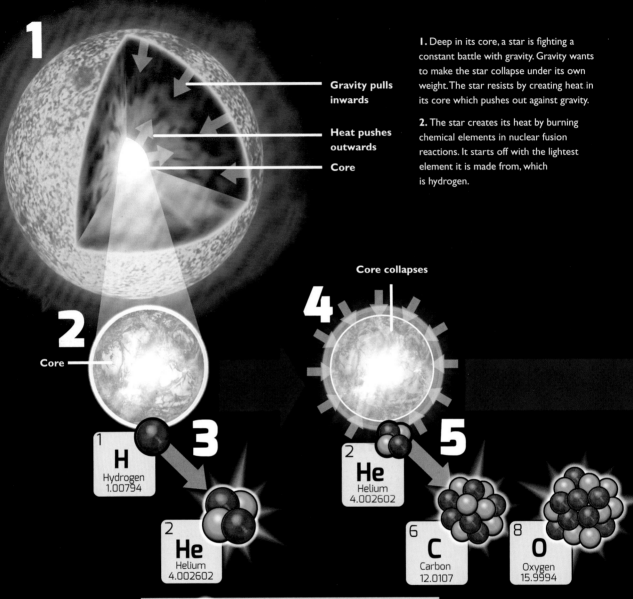

1 Gravity pulls inwards

Heat pushes outwards

Core

1. Deep in its core, a star is fighting a constant battle with gravity. Gravity wants to make the star collapse under its own weight. The star resists by creating heat in its core which pushes out against gravity.

2. The star creates its heat by burning chemical elements in nuclear fusion reactions. It starts off with the lightest element it is made from, which is hydrogen.

2 Core

4 Core collapses

3

1

H
Hydrogen
1.00794

2

He
Helium
4.002602

5

2

He
Helium
4.002602

6

C
Carbon
12.0107

8

O
Oxygen
15.9994

3. Subjected to colossal pressures and 15 million°C temperatures, hydrogen nuclei (protons) are fused together, via the proton-proton cycle, creating the heavier element helium.

7

N
Nitrogen
14.007

Later generation stars that already contain carbon and oxygen (inherited from their stellar ancestors) have an alternative way to liberate energy from hydrogen fusion called the CNO cycle (see page 119). A happy by-product of hydrogen fusion by this method is the creation of nitrogen (an essential ingredient for life).

4. Eventually the star exhausts its supply of hydrogen, and fusion shuts down in the core.

With no more heat being released, gravity gains the upper hand and starts to crush the star's core. As it collapses, the pressure increases within it and temperature rises to 100 million°C.

ASSEMBLING THE CHEMICAL ELEMENTS

Gravity, the great creator, is also the great destroyer – conspiring to crush the very stars it helped to forge. Luckily, the stars have a secret weapon that, for a time at least, can hold gravity at bay and, in the process, create the heavy elements: thermonuclear fusion.

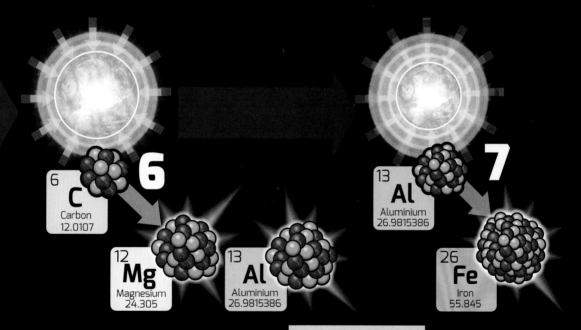

6
C
Carbon
12.0107

6

12
Mg
Magnesium
24.305

13
Al
Aluminium
26.9815386

13
Al
Aluminium
26.9815386

7

26
Fe
Iron
55.845

5. At this temperature, helium fusion begins and the star restabilizes. Helium fusion creates new elements that will be essential when we want to populate our Universe with life – oxygen and carbon.

6. But before very long (about a million years or so) the star runs out of helium to burn as well. With fusion halted once more, gravity takes over again – crushing the core until its temperature reaches 100 million°C and carbon fusion begins – creating heavier elements like sodium and magnesium.

Less massive stars, like our Sun, don't have enough mass to create the pressure needed for carbon fusion, so this is where they die. Fortunately for us, low-mass stars take much, much longer to use up their fuel.

7. The processes of fusion, fuel-exhaustion, core collapse, and re-ignition are repeated (each time forming heavier elements) until, finally, iron is created. Iron is the heaviest element that can be created within a star – to make anything heavier, our star must die.

THE DEATH OF A STAR

A nucleus of iron-56 is the pinnacle of nuclear stability – no other element is as stable. Normally we think of stability as a good thing – we all want a stable relationship, or a stable bridge to walk across – but, for a star, iron's stability spells its doom. Because it is so stable, iron-56 doesn't have the same appetite for alpha particles as its predecessors – so the only way to get an iron-56 nucleus to bond with an alpha particle to make a heavier element is to put more energy into the process than could ever be released from the reaction. So fusion shuts down once more in the core, but this time it won't start up again. The star has run out of options and its fate is sealed.

Even the most massive stars will perish at this point – first collapsing and then exploding in a violent supernova explosion. But only truly massive stars have enough gravitational clout to squeeze the core to this stage: stars of lesser stature will have shuffled off their mortal coils long before now. The Sun, for example, has only enough mass to see it through to the end of helium fusion. Stars of this size end helium fusion by throwing off their gaseous outer layers and spending the rest of eternity as a slowly cooling ball of carbon about the size of Earth called a white dwarf (or, if you are of a capitalist bent, a giant, solar-mass, ten-billion-trillion-trillion-carat diamond).

But even in death, a giant star can contribute to the elemental richness of the Universe – in fact, it is only in the explosive death throes of giant stars that elements heavier than iron, such as gold, lead, mercury, titanium, and uranium, can be produced.

When fusion closes down with the production of iron, the core can no longer support itself and it collapses so violently that the rest of the star's material is caught by surprise and is left hanging above the void like Warner Bros' eponymous animated antihero, Wile E Coyote. Of course, just as it does for the Road Runner's nemesis, gravity eventually catches up with reality and the stellar material falls down towards the core.

At the same moment, the core releases a huge blast wave of gravitational energy that collides with the in-falling material. Where the two fronts smash together, the material is compressed and super-heated – forming a shockwave, where the conditions are so extreme that some nuclei are torn apart and others are bombarded with the resulting barrage of high-energy neutrons. The neutrons are forced to fuse with heavy nuclei – creating elements heavier than iron, most of which are unstable radioactive elements such as uranium (the heaviest naturally produced element), which will decay to become elements like gold.

As the shockwave spreads through the remaining stellar material, the whole kit-and-caboodle is blasted out into space – creating vast clouds of hydrogen, helium,

CONTINUES ON PAGE 123 ➡

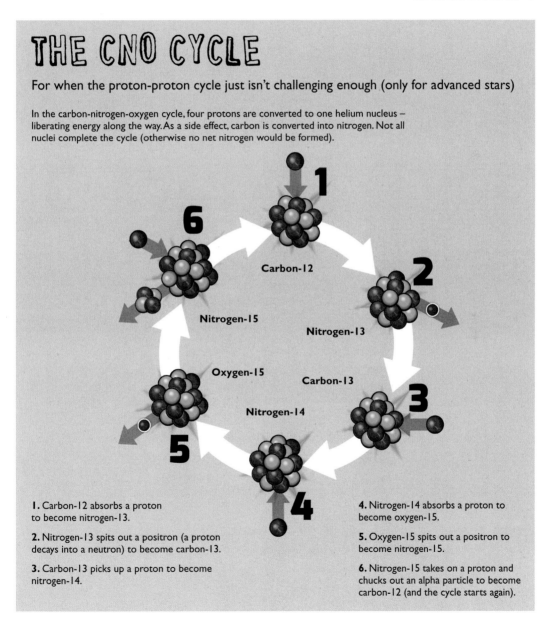

THE CNO CYCLE

For when the proton-proton cycle just isn't challenging enough (only for advanced stars)

In the carbon-nitrogen-oxygen cycle, four protons are converted to one helium nucleus – liberating energy along the way. As a side effect, carbon is converted into nitrogen. Not all nuclei complete the cycle (otherwise no net nitrogen would be formed).

1

Carbon-12

2

Nitrogen-13

6

Nitrogen-15

Oxygen-15

Carbon-13

Nitrogen-14

5

3

4

1. Carbon-12 absorbs a proton to become nitrogen-13.

2. Nitrogen-13 spits out a positron (a proton decays into a neutron) to become carbon-13.

3. Carbon-13 picks up a proton to become nitrogen-14.

4. Nitrogen-14 absorbs a proton to become oxygen-15.

5. Oxygen-15 spits out a positron to become nitrogen-15.

6. Nitrogen-15 takes on a proton and chucks out an alpha particle to become carbon-12 (and the cycle starts again).

DIE HARD: FORGING THE HEAVIEST ELEMENTS

Iron fusion requires more energy to be put in than it releases. Once a star has iron in its core, it is doomed. Fusion reactions shut down and the star is at the mercy of gravity. It is in a star's death throes that the heaviest elements are created.

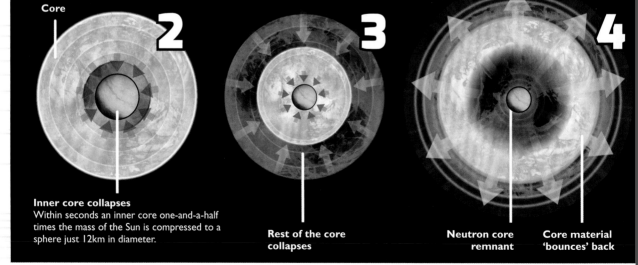

Blue supergiant star

Red giant star
As the star ages, it cools, expands and becomes a red giant.

Core layered with unburned elements

Iron inner core

1. By the time iron is created, a star's core resembles a giant onion – layered with the unburned remnants of all the elements it has created.

2. Unable to fuse iron, fusion shuts down for the last time and the iron inner core violently collapses under its own weight.

3. Moments later, the rest of the core follows suit and all those leftover elements fall towards the collapsing inner core. But the iron core can only collapse so much and, once it has been squeezed into a ball of solid neutrons its collapse halts.

4. The rest of the in-falling core material hits the neutron ball and rebounds – releasing a blast wave of gravitational energy.

Core

Inner core collapses
Within seconds an inner core one-and-a-half times the mass of the Sun is compressed to a sphere just 12km in diameter.

Rest of the core collapses

Neutron core remnant

Core material 'bounces' back

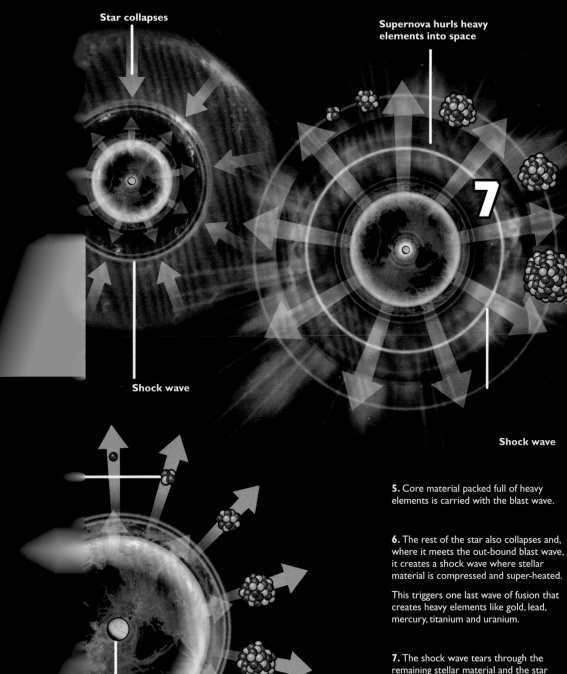

Star collapses

Supernova hurls heavy elements into space

7

Shock wave

Shock wave

Remnant

Blast wave

5. Core material packed full of heavy elements is carried with the blast wave.

6. The rest of the star also collapses and, where it meets the out-bound blast wave, it creates a shock wave where stellar material is compressed and super-heated.

This triggers one last wave of fusion that creates heavy elements like gold, lead, mercury, titanium and uranium.

7. The shock wave tears through the remaining stellar material and the star explodes in a powerful supernova explosion that throws everything the star has made during its lifetime out into space.

But that isn't the end of the story – that leftover ball of neutrons might still have a part to play ...

SUPERNOVA REMNANTS: FROM DEATH, GREAT BEAUTY

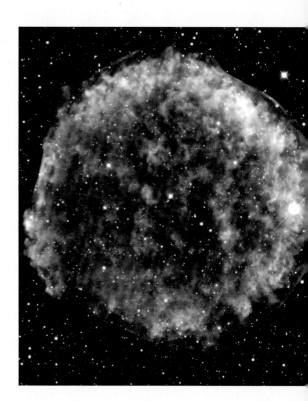

These stunning apparitions are supernova remnants – the scattered remains of an exploded star. Eventually the debris will be recycled to form new stars, enriched with the heavy elements forged in the heart of its predecessors.

Top: Composite image of SN 1572 (Tycho's supernova)

Bottom left: Crab nebula (NGC 1952)

Bottom right: LMC N 49

I'd love to give you evocative names like Agamemnon, Aerope or Derek for these beauties but their names are designed to describe key details. For example, SN 1572 stands for Supernova 1572 (its year of discovery).

carbon, oxygen, iron, gold (and everything else cooked up during the star's lifetime) around the leftover core. These clouds of enriched star stuff are called supernova remnants, or planetary nebulae (a misleading name coined before astronomers understood that they had nothing to do with planet formation). Eventually, these clouds will collapse to form the next generation of stars.

WHEN CORES COLLIDE

Until recently, it was thought that was where the story ended – it seemed to be the only available mechanism for heavy element production. But observations made in 2013 by the mighty Hubble Space Telescope of a burst of energy seen radiating from a galaxy 3.9 billion light years from Earth may have revealed a second possible mechanism for the creation of the heaviest elements: colliding neutron stars.

A neutron star is what is left over when the iron core of a massive star undergoes its final supernova-triggering implosion. In its last seconds, the iron core, which has the mass of between one-and-a-half and three Suns, collapses into a sphere just 12–20km in diameter (that's a ball of matter smaller than the island of Manhattan, which, if it were placed next to the Sun, would exert enough gravitational force to make our local star start orbiting *it!*).

When atoms are that closely packed, all the empty space between the nucleus and its orbiting electrons is squeezed out and the electrons are forced into the protons – converting them into neutrons. Within a matter of seconds, the ball of iron becomes a solid ball of neutrons – it has become a neutron star. As an anecdotal aside, if you were to gather up every one of the 7 billion humans alive today, pack them into an atom-squashing machine (the author isn't aware of such a device but, if it exists, it's probably manufactured by the Acme corporation) and then squeeze them down to neutron star density, all of humanity would fit in a box the size of a sugar cube.

If this stellar remnant (the neutron star, not the human cube) meets another neutron star, their combined gravitational pull means that they feel an irresistible mutual attraction, which – like the affairs of all great star-crossed lovers – can only end in tears.

CONTINUES ON PAGE 127 ➡

FROM OLD CORE TO NEUTRON STAR

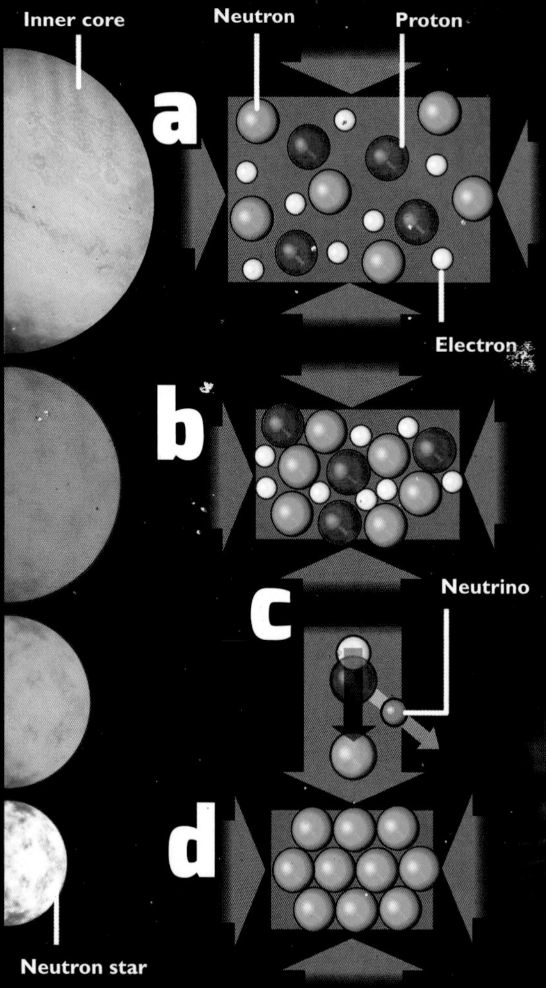

Inner core   **Neutron**   **Proton**

a

Electron

b

Neutrino

c

d

Neutron star

a. Here's the iron inner core just before its final catastrophic collapse. Although dense, the core's atoms are kept at a distance by the electromagnetic force (if protons were the same size as the one on this page, the electrons would be about 30m away).

b. As the core is crushed, the gravitational force overwhelms the electromagnetic force and the atoms are squeezed tightly together.

c. The pressure becomes so extreme that electrons (negative charge) are squeezed into the protons (positive), which emit neutrinos to become neutrons (neutral).

d. The end result is a star made entirely of tightly-packed neutrons, which is basically a ball of solid, super-concentrated matter that can pack the mass of an entire mountain range on Earth into a few square centimetres.

If humanity was squashed in this manner, all seven billion of us would fit in a box the size of a sugar cube.

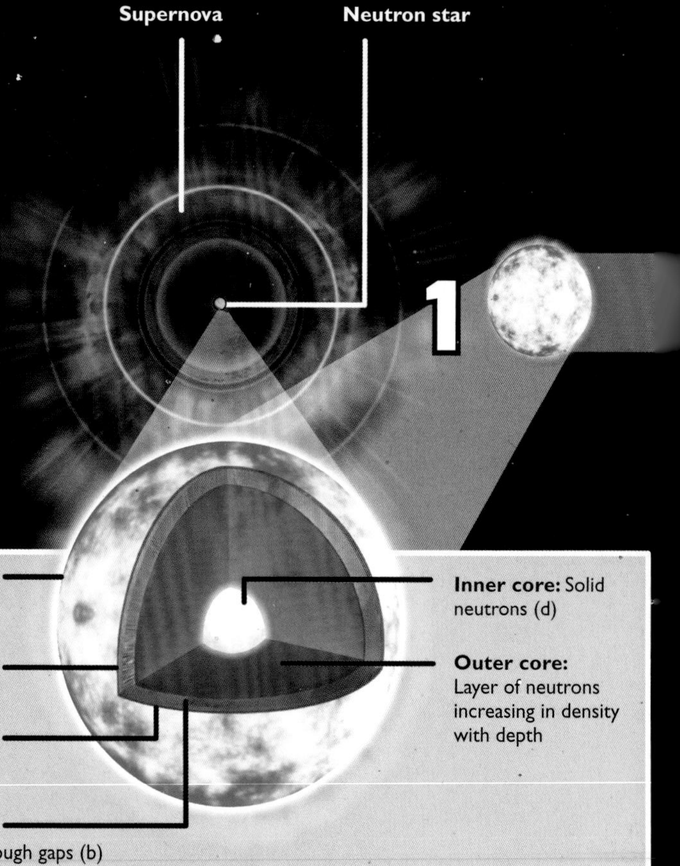

Supernova   **Neutron star**

1

ANATOMY OF A NEUTRON STAR

Atmosphere: A super-dense, super-hot (2 million°C) carbon atmosphere just 10cm thick

Iron envelope: Thin layer of iron atoms

Outer crust: Extremely thin crust made of atomic nuclei and electrons (see (a) above)

Inner crust: Crushed atomic nuclei with electrons flowing through gaps (b)

Inner core: Solid neutrons (d)

Outer core: Layer of neutrons increasing in density with depth

DIE HARDER TEACHING OLD CORES NEW TRICKS

Supernovae may not be the only way to create the heaviest elements. It may be that the vast majority are created by something that supernovae leave behind: super-dense core remnants known as neutron stars.

WHEN NEUTRON STARS COLLIDE

1. A star explodes as a supernova and leaves its neutron star behind. With more than 1.5 Suns-worth of matter squeezed into a 12–20km sphere, neutron stars have humongous gravitational power.

2. So, when two neutron stars stray too close to each other, their mutual gravitational attraction ensures they are drawn together and start to orbit each other.

3. Caught in their ever-decreasing orbit, the (ahem) star-crossed lovers (sorry) move faster and faster until, with each moving at about 67 million mph, they smash together.

4. Some of their material is blasted into space. A shower of neutrons smashes into particles in the surrounding environment – super-heating it and triggering a wave of fusion that creates heavy elements like gold, lead, mercury, titanium and uranium.

2

3

4

Orbiting neutron star   Neutrons   Black hole

5

A neutron star's gravity is so extreme that, if you were to land on one, you would weigh about 7 billion tonnes. But you wouldn't really get the chance to worry about your sudden weight gain because, as you approached the star, you'd be stretched into a piece of human spaghetti and fall towards its surface at more than 4 million mph – where you'd be crushed into a speck of matter smaller than a grain of salt and assimilated into the star's surface.

5. But most of the neutron stars' mass merges in the collision. With so much mass concentrated in one place, the gravitational power becomes overwhelming and the pair collapse into an infinitely small, infinitely dense point from which nothing (not even light) can escape – forming a black hole.

WHEN IS A NEUTRON STAR NOT A NEUTRON STAR?

A rapidly spinning neutron star can emit huge beams of radiation from its poles. These are called 'pulsating stars', or pulsars.

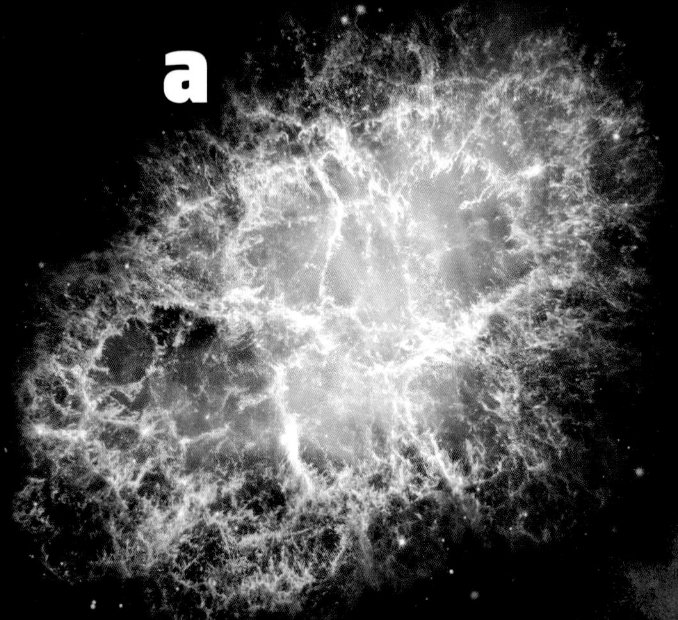

a

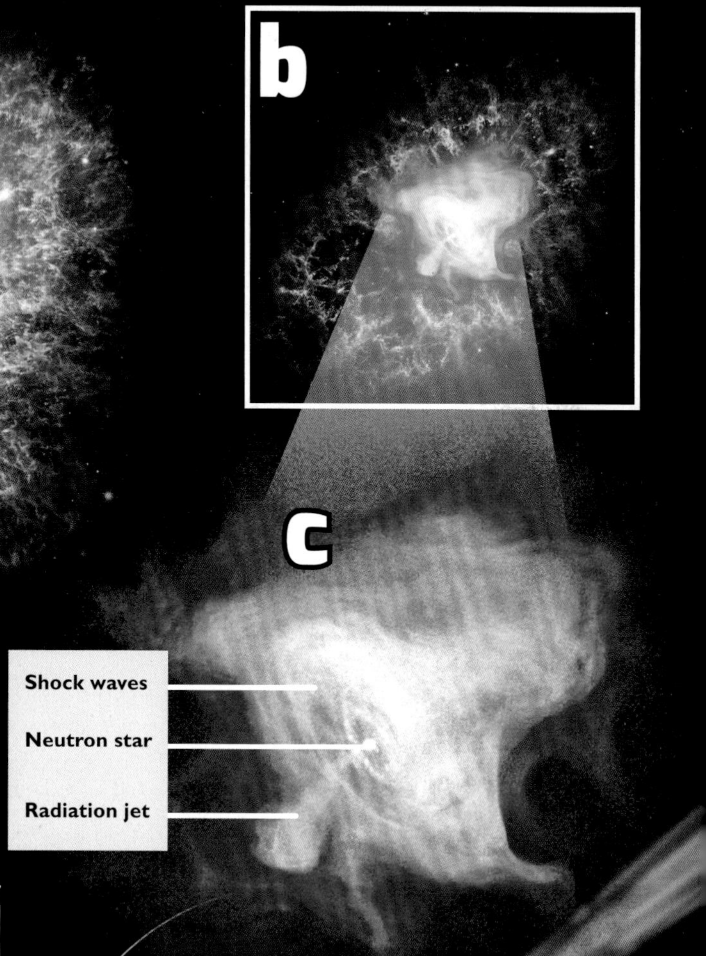

b

c

Shock waves

Neutron star

Radiation jet

Iron core **1**

Neutron star **2**

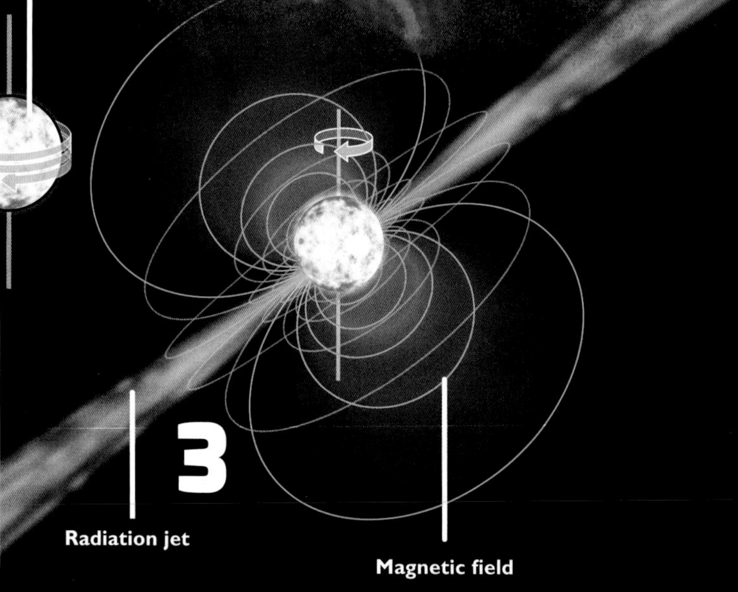

Pulsars were first detected in 1957 by Jocelyn Bell, a PhD student at Cambridge University, when she noticed a regular pulse in the data from the university's radio telescope. Their mysterious mechanical regularity led them to be nicknamed 'LGM' (little green men).

3

Radiation jet

Magnetic field

a. This is the Crab Nebula – a supernova remnant about 6,500 light years away – as it appears in visible light to NASA's Hubble Space Telescope.

b. This is the same nebula with data from NASA's Chandra X-ray Observatory overlaid. It reveals a peculiar structure in the centre of the nebula that is actually the neutron star remnant leftover from the supernova that created the nebula.

c. But this neutron star is actually a rapidly spinning pulsar that is spewing high-energy radiation jets into the surrounding nebula. Where the jets interact with material in the nebula, they create shock waves.

HOW PULSARS WORK

1. An exhausted iron core collapses just before its star explodes as a supernova. Like all stars, this one was spinning before it died and this spin has been preserved in the core.

2. As the core collapses, its spin accelerates, so by the time it has become a neutron star it might be spinning at up to 1,000 revolutions a second (although most only complete a few spins per second).

3. Neutron stars have immensely powerful magnetic fields that, as the star spins, act like colossal electric dynamos – creating powerful electric currents.

These currents flow along the magnetic field lines and work like super-charged particle accelerators – picking up protons and electrons from the surface of the star and firing them into space as beams of high-energy radiation.

The jets are only visible from Earth when they point directly at us and, as the star spins, the jets spin with it. From Earth, we see the jets as pulses of radiations – hence their name 'pulsating stars', or pulsars.

As they orbit each other, they are drawn closer and closer together, and the speed of their orbit increases thanks to conservation of momentum (just as an ice skater's spin speeds up as they pull their arms into their body) until, travelling at a combined speed of more than 130 million miles per hour, they smash together so violently that billions of tonnes of high-energy neutrons are hurled into space. These crash into particles in the surrounding environment – super-heating it and fusing to create heavy elements.

Another by-product of the neutron star collision and the union of all that mass is the creation of a black hole (more on these later) and a blinding flash of radiation, which is called a gamma-ray burst.

Thanks to a clever piece of capitalist-instinct-exploiting PR spin, Hubble's discovery made global news after revealing that as much as 10 lunar masses' worth of gold might be produced from such an event (not immune from such temptations, this author's story for the UK's *Metro* newspaper was accompanied by the headline 'Doubloons from cosmic kabooms' ... I'm not proud of this).

Obviously a single observation does not a scientific consensus make, but, if it turns out that heavy elements are indeed manufactured in such vast quantities by neutron star collisions, it could solve a problem that has been bugging astrophysicists for decades. Because, although the stellar fusion (nucleosynthesis) theory for heavy-element production up to and including iron is pretty, well, iron-clad, when they estimate the amount of heavy elements that could be created in supernova explosions alone, the results invariably fall short of the amounts we see in the Universe today.

Anyway, whether created solely in the explosive death throes of stellar giants or in the collision of neutron stars, or a bit of both, the result is the same: all the elements heavier than lithium are made by stars and are distributed throughout the cosmos by colossal cosmic kabooms.

One day, all those heavy elements will come together to build planets and life to populate (at least one of) them. But a more immediate side effect is that heavy elements have a dramatic cooling effect on interstellar gases (far more so than those first hydrogen molecules). They create dust grains that act as a coolant – radiating heat from the dust clouds and allowing them to collapse faster. As a result, stars start to form more quickly and, because they have less time to collect gases, these second generation stars are smaller, burn less fiercely and so live longer.

So, there we are – after just a few short hundreds of millions of years, our Universe has advanced from being a cloud of simple gases with regions of barely discernible density fluctuations to one populated by massive stars churning heavy chemical elements into the cosmos.

But we aren't finished with our first generation of stars just yet. Much of what is described above is how later generations of stars (already enriched with heavy elements) end their lives. Our first star is made of nothing more complex than lithium (and there's very little of that), which has a nucleus containing three protons and three neutrons, and such stars are generally thought to have been many times more massive than the stars that followed them. Although their size has been downgraded in recent years, they are still believed to have been more than 25 times the mass of our Sun. Stars this massive still manufacture heavy chemical elements (apart from nitrogen, which requires the presence of carbon and oxygen) and still undergo core collapse, explode as supernova and create clouds of enriched gases. However, they are so massive, and their cores contain so much material, that their collapse becomes total and they create one of the Universe's most extreme and enigmatic phenomena: the black hole.

BLACK HOLES: WHEN GRAVITY FLEXES ITS MUSCLES

Stars with cores that contain the mass of more than three Suns don't stop collapsing at the neutron star phase, but, instead, keep on going until all their mass becomes concentrated within a tiny point, called a singularity. And it's the singularity that is the gravity engine that powers a black hole.

It's worth taking a moment to pause and appreciate just how small a singularity is. If you consider that a grain of salt (which is pretty small) might measure in at 0.0001 metres (one preceded by four zeros), to describe the size of a singularity you would have to stick 35 zeros in front of that number one, which looks like this: 0.0000000000000000000000000000000001 metres (remember that that impossibly small speck contains the mass of several Suns squashed up inside it).

When mathematicians attempt to describe what goes on within a singularity, they get numbers that start spiralling into infinities – the laws of physics break down and space and time cease to exist (just as they did in the singularity that birthed the Universe in the Big Bang), so we can't really understand the singularity. But we can understand the effect it has on the space around it.

Around the singularity, gravity is so intense that space-time becomes infinitely curved and creates a gravity well so deep that nothing (not even light) has enough energy to 'climb' out and escape its clutches. The point at which light 'falls' over the edge of the singularity's gravity well and vanishes forever is called the event horizon and it is this dark region that we call a black hole.

Black holes are one of the most evocative, intriguing, and least understood phenomena in the Universe. They have been a staple villain of science fiction for decades and are generally portrayed as sinister behemoths that lurk in the dark recesses of the universe ready to devour anything unfortunate enough to fall within their grasp. But, despite their well deserved destructive reputation, black holes are an essential part of our Universe construction kit – one that'll we'll use in the next chapter to build something truly impressive: a galaxy.

BLACK HOLES: WHEN GRAVITY GOES EXTREME

Not all massive stars finish life as a neutron star: for really massive stars (like our first star) their fate even is more dramatic. Their cores collapse to become the ultimate expression of gravity's power – they become black holes.

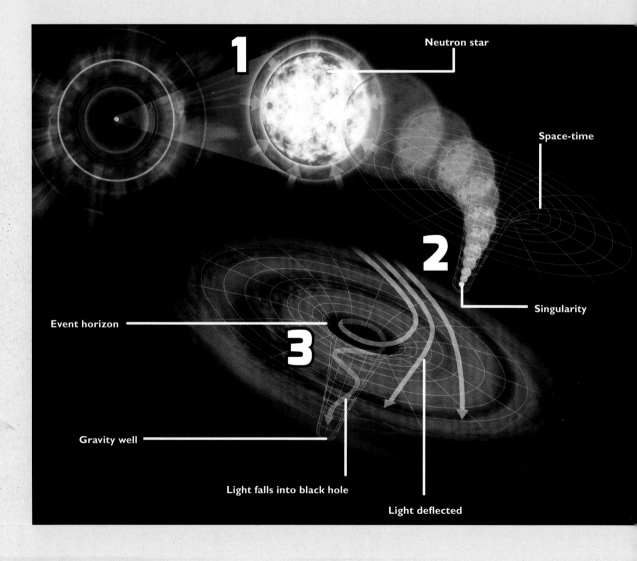

1 Neutron star

Space-time

2

Singularity

Event horizon

3

Gravity well

Light falls into black hole

Light deflected

1. For a star that has a core with more than three solar masses, gravity doesn't stop crushing it at the neutron star stage, but continues until it has been crushed to an inconceivably tiny and incomprehensively massive point, called a singularity.

2. Around the singularity, the laws of physics (as we know them) break down and space and time become infinitely curved. This creates a gravity field so strong that anything that ventures too close is siphoned up and lost forever.

3. The point of no return, where the gravity well becomes inescapably deep, is called the event horizon. Beyond the event horizon not even light can move fast enough to escape – this 'dark' zone is what we call the black hole.

4. Any matter in its vicinity when it forms falls into the black hole. But it can't all get in at once so it 'queues' up at the event horizon – creating a spinning disk of debris called an accretion disk.

5. Of course, space-time isn't a two-dimensional sheet – it is three-dimensional (four if you include time). This means that a black hole is more like a sphere, with a central point where space-time is drawn to a focal point (the singularity) in the centre.

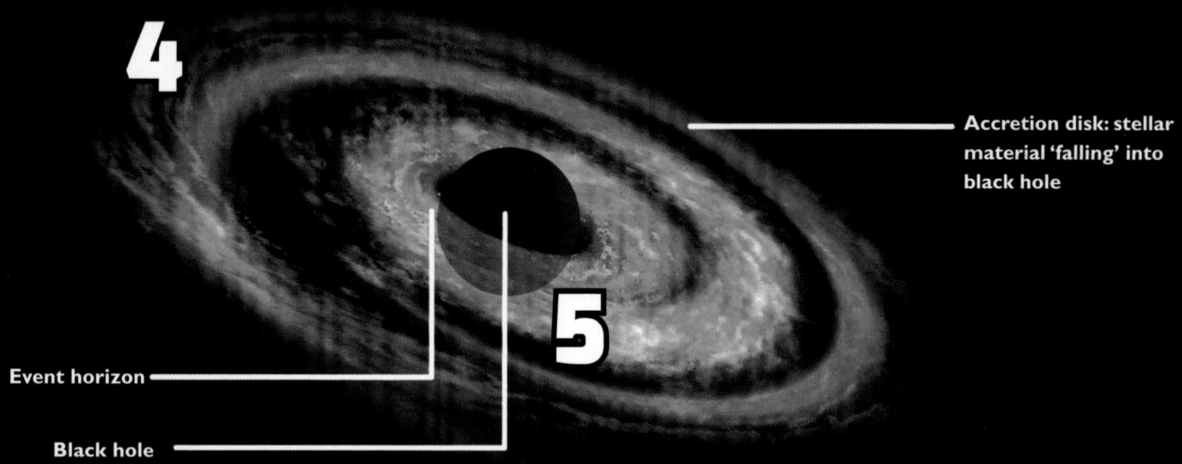

4

5

Accretion disk: stellar material 'falling' into black hole

Event horizon

Black hole

MEET THE GALAXY GARDENERS

In which we unexpectedly (but temporarily) abandon our Universe building site metaphor and switch to a Universe allotment. We grow some large galaxies and employ a black hole to manage our cosmic plot.

So far we have taken a few barren plots of hydrogen gas, planted the seeds of stars, and watched them grow, bloom, die and spread their heavy-element-laden pollen throughout the cosmos – ready to fertilize the next generation of stars.

As every good gardener will tell you, if you want a truly productive garden, you need someone to tend it. The soil needs to be turned, weeded and fertilized. Plants that have grown too close together need to be thinned and their growth managed. In short, you need a gardener and, as it turns out, in our new galactic garden metaphor, that role might – surprisingly – be being filled by a supermassive black hole.

Until recently no one suspected that the growth of galaxies and the stars they contain was governed by anything other than the (not so) simple laws of physics. But new research into the link between black holes and galaxies is revealing that they might share a much more intimate relationship and that their evolution, lives and, ultimately, their fates are inexorably connected.

THE BLACK HOLE COMETH

The idea that something as bizarre as a black hole might exist at all emerged uninvited from the pages of Einstein's theory of General Relativity (like a forgotten gas bill once used as a page marker, dropping from a book). When the great physicist redefined gravity as being an effect caused by massive objects distorting the fabric of space-time, he inadvertently allowed for the existence of something truly terrifying – objects with so much mass that they distort the fabric of the Universe to such an extent they could wrap themselves away in a sort of space-time cocoon that not even light could escape, and in which space and time cease to exist.

Big Bang	Particles form	CMB	Dark ages (first dark matter structures)	First stars and active galaxies
13.82bn years ago		377,000 years after Big Bang		200 million years

Einstein himself hated the idea of black holes and was embarrassed by their theoretical existence, believing that they revealed a flaw in his calculations. Yet exist they did. After decades of speculation, they left their theoretical existence behind in the 1970s and emerged as one of the most enigmatic phenomena in the cosmos.

Today, we know that black holes are common throughout the Universe and that every galaxy (even our own) is home to millions of small black holes with the mass of a few Suns. They wander around interstellar space, snacking on gas, dust, and the occasional errant star. But we also know that hiding at the heart of every large galaxy there is something even more extreme – a supermassive black hole containing the mass of millions or billions of Suns that seems to act like a hub around which the rest of the galaxy slowly rotates.

Astronomers suspect that there must be a link between a galaxy and the supermassive black hole at its heart. The exact nature of that connection is far from being fully understood. Did they evolve in tandem? Did the black hole form as a by-product of the galaxy that contains it; or did the galaxy itself form around the black hole as vast clouds of gas and stars were drawn towards its massive gravitational influence? It really is a 'chicken or the egg' situation.

Unfortunately, black holes and galaxy formation are two of the least understood phenomena in the Universe but there are enough clues to allow cosmologists to start building a picture that seems to show that supermassive black holes do indeed influence the evolution of their host galaxy.

But how could a black hole possibly affect something as big as a galaxy in any sort of meaningful way? Even the most massive are dwarfed by the galaxy that contains them. For example, a supermassive black hole with the mass of billions of Suns represents only a tiny fraction of the total mass of the galaxy it lives in – usually less than one per cent. That's like saying that an object the size of a boulder could influence an entire planet the size of Earth.

But we are getting ahead of ourselves somewhat. Leaving aside the gardening metaphor, before we look for a link between supermassive black holes and galaxies, we first need to build them.

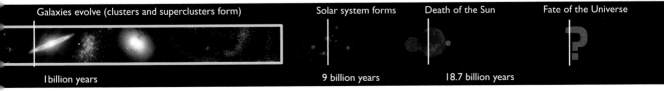

Galaxies evolve (clusters and superclusters form)   Solar system forms   Death of the Sun   Fate of the Universe

1 billion years   9 billion years   18.7 billion years

BUILDING A GALAXY

○○○

About 700 million years have passed since everything kicked off in the Big Bang and we find ourselves in a Universe populated with lots of protogalaxies – those fuzzy little clouds of gas and dust populated by stars orbiting around a centre of mass – and it's all doing a very fine job of reionizing the cosmos and bringing the dark ages to an end.

Now, bear in mind that at this point the Universe isn't really all that old and it is still a long way off from being the size it is today. This means there is lot less 'empty' space for all those protogalaxies to swim around in and eventually they are going to start bumping into each other.

Like all massive objects, when two galaxies stray too close, they find themselves ensnared by their mutual gravitational attraction (they roll down the gravity hill) and start orbiting each other. Caught in an inescapable circular dance, the pair spiral together until, in a swirling maelstrom of gas and stars, they collide.

But this isn't the sort of cataclysmic collision that might occur when two planets smash together; instead, because everything within the protogalaxies is so thinly spread, they just merge – their gases intermingling until the stars suddenly find themselves living in a slightly bigger piece of galactic real estate.

For small galaxies, these mergers are still quite disruptive and their material becomes all churned up as their combined mass struggles to find a common centre of gravity. Although their stars are never in much danger of colliding, giant clouds of molecular hydrogen within the two galaxies are and, when these do smash together (in so far as diffuse clouds of gas can 'smash' together) they condense – sparking a flurry of star formation.

Nor do their existing stars have it easy. In the collisions, they find themselves torn from their orderly rotations and they are flung into random orbits within the newly-merged galaxy. If they are lucky, they will find new orbits around a new centre of mass as the galaxy settles down and resumes the slow rotation that it inherited from the initial collapse of the primeval hydrogen clouds.

But, the unlucky ones might find themselves hurled out of the galaxy altogether – doomed to spend the rest of their days wandering the cosmos alone as stellar outcasts, known as rogue stars.

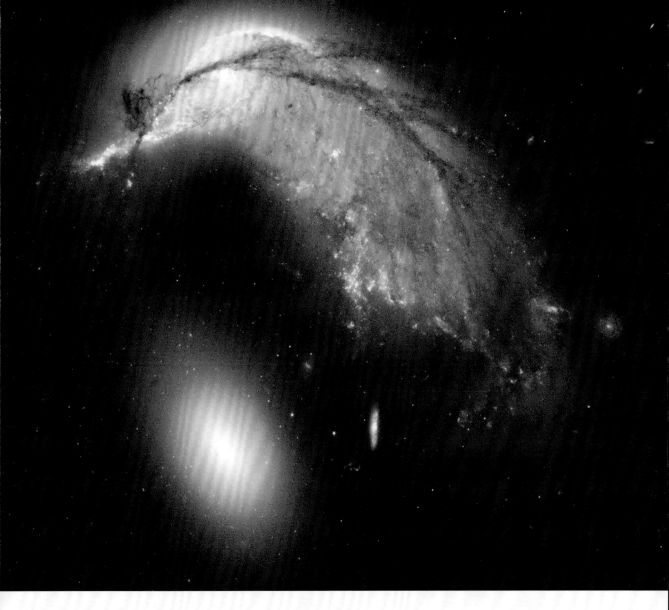

This splendid image from NASA's Hubble Space Telescope shows an interacting pair of galaxies known as Arp 142. The Penguin, or NGC 2936, was once a neat spiral galaxy but its material has been scrambled and stretched by its interaction with the small elliptical galaxy NGC 2937. The Penguin's blue plumage is actually regions of intense star formation triggered by the interaction. Behold the cosmic Penguin.

GROWING MASSIVE GALAXIES

Just as the first stars and protogalaxies were formed from particles brought together by their mutual gravitational attraction, so the giant galactic structures we see today were formed by small galaxies drawn together in the same way.

1. This is a picture of galaxy formation in action. At the centre of the image is a cluster of dwarf galaxies that are being drawn together. Over millions of years, they will merge to create a new galaxy. Called the Spider's Web galaxy, the cluster is 10.6 billion light years away, so we are seeing it as it appeared just three billion years after the Big Bang.

The merging of two dwarf galaxies is like an intricate ballet where the dancers slowly become intertwined over the course of millions of years.

2. Two dwarf galaxies become gravitationally bound and begin slowly dancing around their shared centre of mass.

3. As they move closer, they extend tenuous arms of gas and dust, which wrap around their partner's core.

4. Eventually the pair move into a close embrace – their stars and gases mingling as their combined spin whips their material into a swirling frenzy.

5. Their cores merge and the pair becomes one. Gas and dust fall to the centre of the newly-formed galaxy – sparking bursts of star formation.

6. The new heavy-weight galaxy draws more small galaxies towards it. With each merger, the galaxy gains mass and new waves of star formation are triggered.

7. Eventually the galaxy gains enough mass that its structure is hardly disturbed by the absorption of dwarf galaxies.

HOW GALAXIES GET THEIR HEART

This simulation of a galaxy merger shows how gas and dust 'falls' inwards and settles at the galaxy's centre. Over time, spin causes the cloud to flatten out – creating the familiar galactic disk.

GALAXY MERGERS IN ACTION

Above: These two galaxies are about 300 million light years away (so we are seeing them as they appeared before the rise of the dinosaurs). The group, known as Arp 273, have been distorted by their mutual gravitational pull. The smaller galaxy shows signs of active star formation, which suggests it might have passed through the larger galaxy (demonstrating the insubstantial nature of a galaxy).

Below: A more evenly matched pair – the two galaxies known as The Mice will merge to create a spiral galaxy in about 400 million years (although they are 290 million light years away, so they'll merge in about 110 million years ... we just won't see it for 290 million years after that).

GALAXY CLASSIFICATIONS

Astronomers still use a system devised by Edwin Hubble in 1926, which sorts galaxies into categories based on their appearance.

Spiral

Spiral galaxies are perhaps the most iconic galaxy shape. They are flattened disks of rotating gas, dust, and stars.

Like huge cosmic cartwheels, they consist of a dense central hub of old stars around which rotate spiralling arms – usually sites of star formation. The Milky Way is a spiral galaxy.

Elliptical

These look like fuzzy eggs or footballs and contain very little gas from which new stars can form, so most are old and sterile. They are thought to be the result of violent galactic collisions that ejected the majority of their star-forming gases. The very oldest stars in the Universe are found in elliptical galaxies.

Lenticular

Somewhere between spirals and elliptical, these are disk galaxies with a dense galactic centre, like a spiral, but they don't have the characteristic spiral arms. Also, they have lost most of their interstellar gases, so there is very little active star formation and they consist of mainly very old stars, like an elliptical.

Irregular

As their name suggests, irregular galaxies have no definable structure. They might be the results of collisions or mergers and haven't had the chance to settle down into a defined shape, or they might just lack the rotational energy required to form spirals or ellipticals.

SUBCLASSIFICATIONS

Galaxies are subdivided into different categories a to variations in their shape:

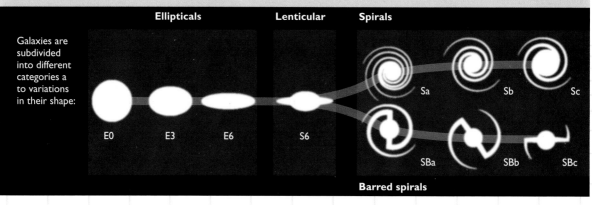

Ellipticals · Lenticular · Spirals

E0   E3   E6   S6   Sa   Sb   Sc

SBa   SBb   SBc

Barred spirals

This process is repeated and replayed as the galaxy continues its cosmic wanderings – slowing accumulating mass as it assimilates more and more protogalaxies into its growing bulk. As it gains mass, the galaxy starts spinning faster and faster and it starts to flatten out to form a disk (like a lump of pizza dough sent spinning into the air). In just a few short billions of years, a protogalaxy that started out as little more than a well-lit cloud of gas finds that it is a full-grown galaxy with billions of stars spread out over eight trillion cubic light years of space – all slowly turning around a central hub like a colossal, sparkly cartwheel.

One day, in a far off distant future, in a land known as Greece, a man gazing at the heavens will see a fuzzy, milky band of light that seems to span the heavens and call it *galaxias kyklos*, or 'milky circle'. Later still, it will be called the Milky Way and the Greek word for milky, *galaxias*, will provide the name for all such structures – the galaxies.

The ancient Greeks were ill-equipped to appreciate the true nature of the Milky Way (it will take until the 17th century and telescopic peerings of Galileo Galilei to reveal that the 'milk' is actually made up of countless stars), but they would have appreciated the almost mythical beast that lurks at its centre – a terrifying creature capable of devouring stars and bending entire galaxies to its will: the cosmic Kraken known as a supermassive black hole.

HOW OLD IS THE MILKY WAY?

It is thought that our home galaxy had pretty much assumed its present spiral disk form by the time the Universe was between 3 and 4 billion years old. But the oldest stars found in the Milky Way were formed just 200 million years after the Big Bang, which makes sense when you consider it was built up from many small galaxies.

BUILDING A SUPERMASSIVE BLACK HOLE

Science is still trying to figure out the mechanisms that drive the formation of supermassive black holes, but there are a few leading theories. These favour a top-down process in which black holes came first and the stars and galaxies followed, or a bottom-up process in which stars live, die and create black holes (see also pages 144–5).

One bottom-up theory is that, during galaxy mergers, turbulence might whip the combining materials into giant whirlpools of swirling gases (like an oar passing through water creates vortices in its wake) that concentrate the gas at its centre. This sudden conglomeration of mass sinks to the galaxy's centre, collapses under its own weight and forms a sort of ultra-hyper-megamassive star the size of tens of thousands of Suns. In a fraction of a second, the core, which just can't create enough heat to hold back all that mass, collapses to form a black hole. Because it bypassed the whole messy supernova bit, the young black hole will have 10,000 stellar masses worth of material right on its doorstep to feast on – before moving on to devour anything else that strays too close.

Another theory, which takes a top-down approach, suggests that supermassive black holes can trace their history back to the very beginnings of the Universe and might have had a hand in creating the galaxies themselves. The idea is that, when dark matter first settled into its network of filaments, at the nodes where the filaments joined, they created gravity wells deep enough to suck in huge quantities of normal matter. This then collapsed to make something very similar to the ultra-hyper-megamassive star above, which, in turn, collapsed to become a supermassive black hole.

These black holes would have then behaved like a fisherman's hook – attracting and ensnaring the clouds of normal matter and assisting their collapse into protogalaxies. Huge outflows of energy from the black holes might then have kick-started the regions of localized collapse that then led to the formation of the first stars.

But perhaps the most promising theoretical mechanism (and archetypal bottom-up process) shows the black holes evolving in tandem with their host galaxy.

A GALACTIC GET-TOGETHER

Let's rewind the clock a few billion years, regress our galaxy and return it to its original protogalaxy form. If we zoom in and take a closer look, aside from a few stars and lots of gas, you won't see very much at all because black holes by their very nature are really quite black, but if we don a pair of my Black Hole Revealifying Goggles (patent pending), you will see that the protogalaxy is peppered with small black holes left behind by expired stars.

Most of these aren't very big, comprising just a few solar masses, but some are a little bit more massive. These might have been born from stars that formed in binary or trinary systems. Instead of having to subside on the remains of just their parent star, these lucky little blighters have a sibling or two to feed on – growing from just 10 solar masses, for example, to 20 or 30 solar masses in just a single bout of stellar fratricide.

Because the engorged black hole is significantly more massive than its surroundings, it might sink towards the centre of the galaxy (like a stone sinking through a vat of porridge), snacking on the odd gas cloud as it goes, and once there will face an eternity of enforced fasting. But, this one is in luck because his galaxy is just about to merge with another.

The two galaxies collide, get all mixed up and then settle down as a larger galaxy but, within the gaseous confusion, our black hole has met a pal. His friend, who came from the other galaxy, is also a little more massive than his galactic compatriots and the two find themselves drawn together and, like a miniaturized version of the spiralling waltz performed by their home galaxies, the two black holes merge, combining their mass to become a single, more imposing gravitational presence.

With its mass increased, the black hole dutifully sinks to the centre of the new galaxy, snacking on all the newly injected gases along the way before settling down to await the next galaxy-delivered payload of food.

And so, as more galaxies merge, the process of merging, sinking, feeding and merging is repeated until the black hole has become a supermassive beast, a cosmic tumour with the mass of millions of stars, pulsing away deep within the heart of a full-grown galaxy.

THE BLACK HOLE POWERHOUSE

Now that we have a black hole that packs the gravitational punch of millions of stars, we can move on to the next earth-shattering revelation ... (I'll give you a moment to sit down) ... not all black holes are black! I realize that's a statement that seems to fly in the face of accepted wisdom (in fact, there's a great deal about black holes that just doesn't fly in the face of wisdom, but laughs in its face, slaps it hard across the cheeks, insults its momma and steals its lunch money), but it's a crucial quirk that might have significant ramifications for the rest of the galaxy.

CONTINUES ON PAGE 146 ➡

PUTTING THE 'SUPERMASSIVE' INT

Supermassive black holes with the mass of hundreds of millions of Suns are thought to live at the centre of almost every galaxy. We don't know for certain how supermassive black holes first gathered their extraordinary mass, but there are three leading theories ...

THREE ROUTES TO BLACK HOLE OBESITY

The first two theories create unstable stars with the mass of tens of thousands of Suns.

Core collapses

Theory A: Dark matter collapse

Before even the first protogalaxies condensed from the primordial hydrogen fog, dark matter was making its gravitational presence known as a vast network of filaments.

Where the filaments joined to form junctions, or nodes, the sheer weight of dark matter might have been enough to draw in huge quantities of matter – creating an immense star (or stellar giant) with the mass of tens of thousands of Suns (similar to that shown in the picture above).

Theory B: Galaxy collisions

This theory would take place hundreds of millions of years after that of the dark matter collapse theory, during the period of galaxy evolution ... When two galaxies collide the impact could create powerful zones of turbulence (similar to those shown above) that would whip gases into a vortex. This could then suck up huge amounts of gas, which would sink to the centre of the galaxy and collapse to form a hugely unstable, hugely huge star (or stellar giant).

Whether the stellar giant is created following the process of either theory 'a' or theory 'b', the result is the same – its astonishing mass makes it impossible for fusion within the core to support its bulk. So the core almost instantly collapses to create a massive black hole at the centre of the star. This then gains mass by devouring the remains of its parent star and snacking on passing stars and dust clouds.

Black hole

Radiation jets

Incredibly, this process does take place in the Universe today (albeit on a smaller scale) within the cores of the most massive stars in the cosmos. Wolf–Rayet stars are so massive that, instead of exploding as supernovae, their cores collapse within the star – leaving the star with a black hole heart that eats the star from the inside.

BLACK HOLES

Theory C: Galaxy mergers and large meals

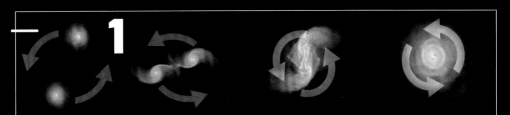

Black holes meet and merge

Accretion disk

2

Star being accreted into black hole

The third theory is that when two galaxies merge, their black holes sink to the centre of the new larger galaxy. They become gravitationally bound, orbit each other, and eventually merge to become a more massive black hole (1). This process could be repeated after each galaxy merger – creating ever more massive black holes.

During each galaxy merger, stars within the colliding galaxies are thrown out of their orbits. Any that stray too close to the central black hole will be stripped of their gases. These are added to the accretion disk, eventually finding their way into the black hole (2).

Over millions of years, the black hole can accumulate the mass equivalent of hundreds of millions of Suns – becoming a supermassive black hole.

Radiation jet

Supermassive black hole-powered quasar

Surrounded by vast clouds of gas and dust to feed on, the supermassive black hole can radiate hundreds of millions of stars' worth of radiation. Galaxies with such active black holes at their centre are called quasars (3).

As the galaxy ages the black hole runs out of food and eventually becomes dormant – until the next meal strays into its lair.

Because star formation was slower in the early universe, there was a lot more gas floating around to provide food for supermassive black holes than there is today.

BLACK HOLES DON'T SUCK

It is common to hear that black holes 'suck' up everything that passes too close – as if they are some sort of super-sized cosmic vacuum cleaner. But there is no suction involved when matter is swept into a black hole. Instead, black holes are so gravitationally intense that they wrap themselves up in the fabric of space-time and drag it inwards. Anything travelling through that particular chunk of space is dragged along with it until, at its closest point, space-time is flowing inwards so quickly that nothing can move fast enough to swim against it.

True, in the region beyond the event horizon (or 'edge'), where light and matter become eternal prisoners of gravity, they are as black as black can be. But, at the same time, black holes can radiate enough energy to outshine an entire galaxy of stars and it's all down to the extraordinary power of the forces that converge at the event horizon.

A black hole will swallow anything that strays too close, but, like water draining from a sink, there is only so much that can flow down the plug hole in one go. So, although the black hole can strip the gases from an entire star, it can't swallow it all at once and the excess stellar material builds up around it to create a spinning disk of material (like water circling the plug hole) called the accretion disk.

In the orbiting accretion disk, the outer regions move more slowly than those closer to the black hole – think of a bicycle wheel with some of those clacker things attached to the spokes (the sort you used to get in cereal packets in the 1980s). If you put one at the far edge near the tyre and another near the hub, the outer clacker has to cover a greater distance during the rotation so it moves faster through space. This is all well and good for a rigid bicycle wheel, but a black hole's accretion disk is a fluid, dynamic system, so different bands of material will rotate at different speeds. Where the different bands interact, the particles rub together, and movement, or kinetic, energy is transferred into heat. Because they lose movement energy through friction, the particles slow down, which, as you might recall, makes them more susceptible to the pull of gravity (they lose orbital energy), so the material spirals inwards and falls towards the black hole.

HOW BIG IS THE EVENT HORIZON?

Even the most massive black holes have event horizons that are dwarfed by the galaxy they inhabit.

The event horizon for a 10 solar mass black hole would be about 37 miles across – the same width as America's smallest state, Rhode Island, or the narrow bit on the boot of Italy.

For a black hole with a mass of 100,000 Suns, the event horizon would be about 370,000 miles across – about the same as the four Jupiters.

A 4 million solar mass black hole would be about 15 million miles across and would slot neatly within the orbit of Mercury.

EVENTS AT THE EVENT HORIZON

The point at which material 'falls' over the lip of the black hole – the point of no return, where even the speed of light isn't fast enough to escape – is called the event horizon, or the Schwarzschild radius (after the German astronomer Karl Schwarzschild who proposed its existence in 1915). It is at the event horizon that things really kick off.

Like planets, galaxies and pretty much everything in the universe, black holes spin – it's just that black holes spin very quickly indeed. Black holes are formed from stars and all stars rotate. So, when a star's core starts to collapse, conservation of angular momentum causes the star's rotation to speed up (think of a spinning ice-skater – yes I'm using this tired old metaphorical chestnut again – tucking his arms in to speed up the spin). By the time it has collapsed to a neutron star, it can be spinning as fast as 1,000 times a second so, by the time it has collapsed to a singularity, a black hole might be spinning at a good percentage of the speed of light. Although, it's more accurate to say that it is space-time itself that is spinning at close to the speed of light.

CONTINUES ON PAGE 151 ➡

HOW TO MAKE A BLACK HOLE SHINE

Black holes are more than just holes in the fabric of space and time that sit around siphoning up light and matter. They are also gravity-powered engines with the ability to extract energy from matter in ways that make the nuclear furnaces of stars look inefficient.

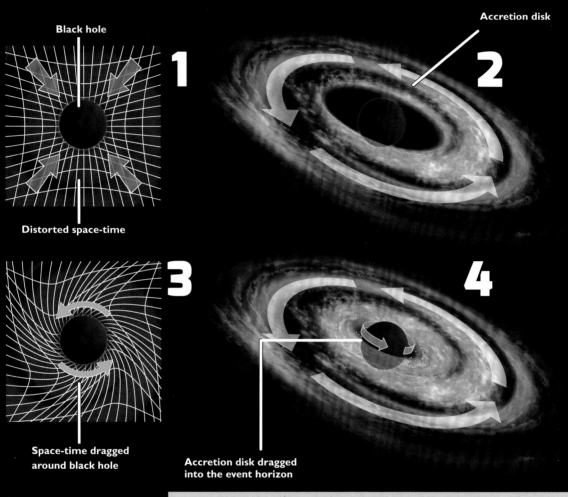

Black hole

1

Distorted space-time

Accretion disk

2

3

Space-time dragged around black hole

4

Accretion disk dragged into the event horizon

1. Here is our black hole. It bends the fabric of the Universe around it – making a gravitational dent so deep that not even light can escape it.

A black hole is the collapsed remnant of a dead star's core. It is in a star's nature to spin and, when it dies, this spin is transferred to its core. As it collapses under the weight of its own gravity, the core's spin accelerates until, by the time it has become a black hole, it can be spinning at almost the speed of light.

2. Around it is an accretion disk – a swirling dish of gas and dust that builds up around the black hole. If the black hole is just chilling out, the orbital momentum of the material in the disk stops it from falling in (like the way the Earth doesn't just fall into the Sun).

3. As a black hole spins, it drags the fabric of the Universe (space-time) around with it. Space itself gets all twisted up around it, like a sheet caught in spinning drill bit – a process known as frame-dragging.

4. As space-time is pulled inwards, the material in the disk is dragged closer. It is when the material reaches the event horizon that things really kick off. This is the point where the black hole's gravity becomes so extreme that not even light can escape.

Past the event horizon, space-time is falling into the black hole faster than the speed of light – so anything else in that portion of space is also accelerated to faster than the speed of light (which is why light can't escape – space is flowing inwards faster than light can move outwards).

5   Black hole

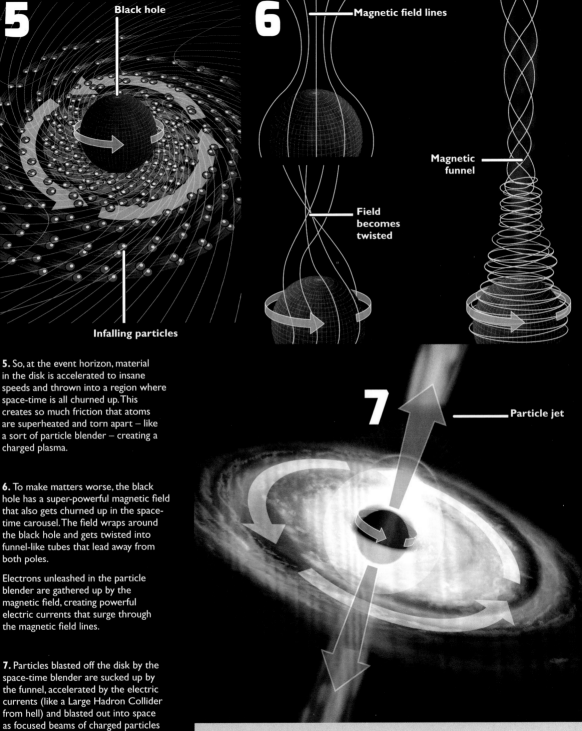

Infalling particles

6   Magnetic field lines

Field becomes twisted

Magnetic funnel

7   Particle jet

5. So, at the event horizon, material in the disk is accelerated to insane speeds and thrown into a region where space-time is all churned up. This creates so much friction that atoms are superheated and torn apart – like a sort of particle blender – creating a charged plasma.

6. To make matters worse, the black hole has a super-powerful magnetic field that also gets churned up in the space-time carousel. The field wraps around the black hole and gets twisted into funnel-like tubes that lead away from both poles.

Electrons unleashed in the particle blender are gathered up by the magnetic field, creating powerful electric currents that surge through the magnetic field lines.

7. Particles blasted off the disk by the space-time blender are sucked up by the funnel, accelerated by the electric currents (like a Large Hadron Collider from hell) and blasted out into space as focused beams of charged particles and radiation.

A black hole can liberate about 28 per cent of the infalling matter's mass into energy – a process almost 50 times more efficient than nuclear fusion.

By this process, a black hole can emit more energy than a million Suns, but these are cosmic wimps when compared to their supermassive cousins ...

BENDING SPACE AND TRAVELLING IN TIME

HOW BLACK HOLES COULD BE USED AS TIME MACHINES

Black holes have powerful effects on the fabric of space-time that surrounds them – effects that could, in theory, be exploited to travel into the future.

TIME IS RELATIVE

Einstein showed us that the rate at which time passes depends very much on where you are and what you are doing – time is strangely malleable and elastic. In the end, it's all relative.

'When a man sits with a pretty girl for an hour, it seems like a minute. But let him sit on a hot stove for a minute and it's longer than any hour. That's relativity.'

ALBERT EINSTEIN

What Einstein is saying is that, far from being universal, time is actually very personal – how we experience time depends on who is measuring it.

His theory of General Relativity revealed that, rather than being an arbitrary and intangible method of charting our lives, time is actually woven into the fabric of space itself – time is the fourth dimension.

Einstein's discovery showed that, if you can manipulate space-time, you can manipulate time itself. The downside is that is not easy for us to manipulate space-time – luckily black holes are custom-made to do just this.

USING GRAVITY TO TRAVEL IN TIME

1. We already know that massive objects like stars distort space-time around them, creating gravity wells – the greater the mass of the object, the greater the distortion. We also know that gravity obeys the inverse-square law – getting stronger as you approach the object – meaning that the closer you get to its centre of gravity, the more space-time becomes stretched.

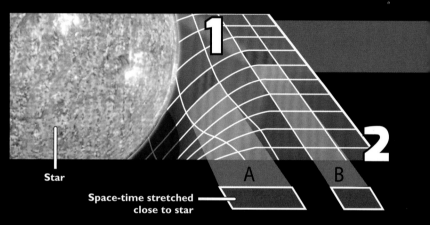

Star

Space-time stretched close to star

A   B

2. This stretching of space also stretches time. Time passes slower for someone in square A compared to someone in square B.

BUT WHY DOES STRETCHING SPACE EQUAL SLOWER TIME?

Einstein's equations tell us that the speed of light is constant relative to the observer. No matter where you are, you will always see light travelling at 300,000 kilometres per second.

a. Observer A has to see light travelling at the same speed as observer B. But in the stretched space of square A, light has further to travel.

b. Since light can't travel faster than 300,000 kilometres per second, the only way it can have enough time to cover the distance is if time moves more slowly in square A.

c. If observer B were to peer into square A, it will appear as if time is moving slower in square A – even though, from their point of view, each observer experiences time flowing at 'normal speed'.

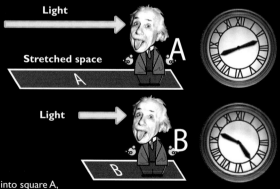

Light

Stretched space

A

Light

B

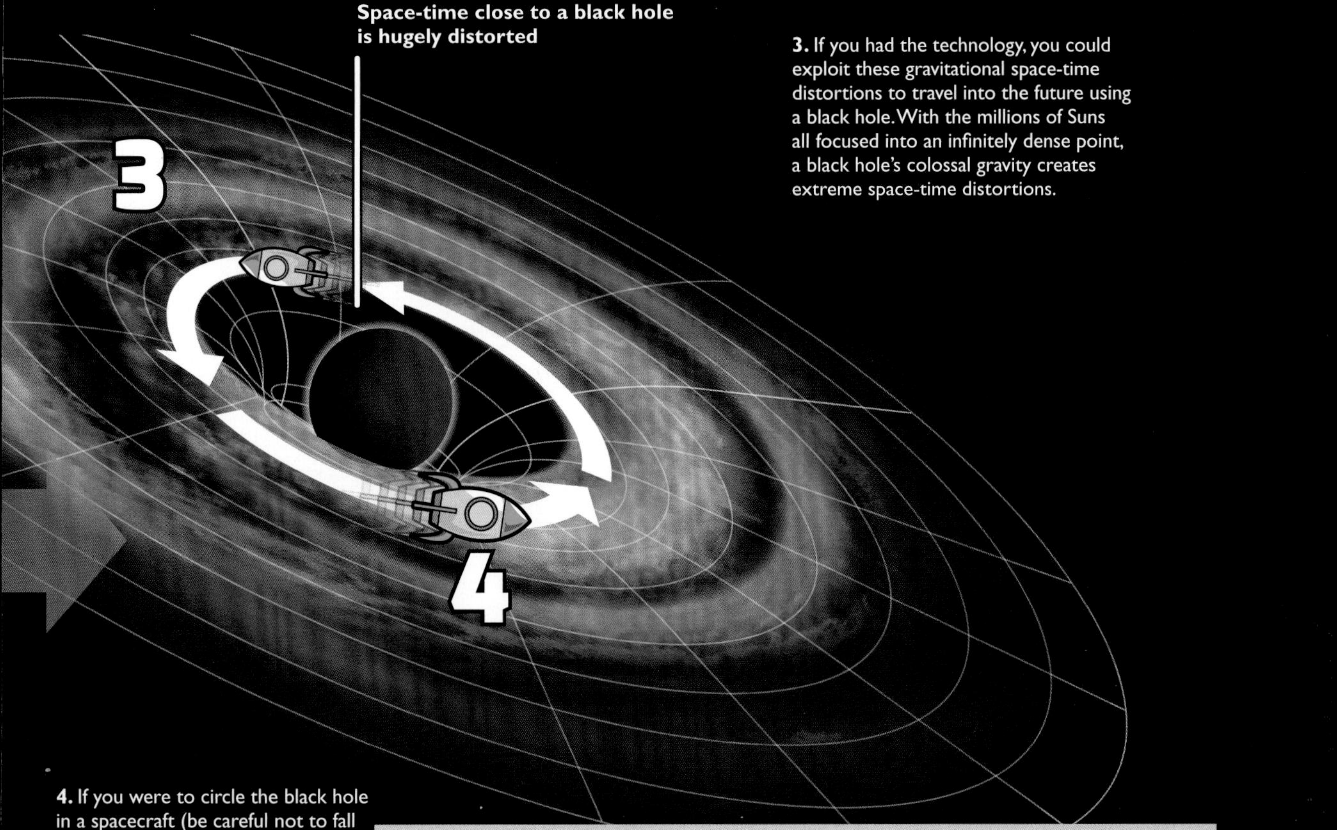

3. If you had the technology, you could exploit these gravitational space-time distortions to travel into the future using a black hole. With the millions of Suns all focused into an infinitely dense point, a black hole's colossal gravity creates extreme space-time distortions.

3

4

4. If you were to circle the black hole in a spacecraft (be careful not to fall in) for five years and then return to Earth, you would find that a decade or two has passed on Earth – you will have travelled into Earth's future.

This does work in practice (albeit on a less dramatic scale) with GPS satellites. Because they travel through space slightly further away from Earth's centre of gravity, they experience time moving slightly faster than we do on the surface – gaining a third-of-a-billionth of second a day.

As it spins, the black hole gets 'snagged' on the fabric of space-time that surrounds it and time and space get all twisted up around the black hole, like a sheet caught in a spinning drill bit, in a process astronomers call frame dragging.

At the event horizon, the combination of extreme gravitational energy, friction and turbulence acts like a particle blender – heating the disk material to hundreds of thousands of degrees, tearing apart its atoms and accelerating them to almost the speed of light.

To make matters worse, the rapidly spinning magnetic fields that surround the black hole act like an enormous electric dynamo, creating powerful electric currents that flow along the lines of magnetism and behave like a Large Hadron Collider from hell, accelerating the particles to within a whisker of the speed of light.

Above and below the black hole, the magnetic field is twisted into funnel-like tubes that lead away from both poles. Those particles still on the 'escape' side of the event horizon are so energized that they emit huge quantities of energy. Together they are channelled by the black hole's powerful magnetic field and thrown out into space as jets of hyper-accelerated charged particles and intense radiation that blast into space at close to the speed of light.

This process (the most efficient matter-to-energy converter in the Universe) means that a black hole can emit more energy than millions, or even billions, of Suns.

QUASARS

Galaxies that have such an active supermassive black hole at their centre are called quasars. The name is a contraction of quasi-stellar objects because, when they were first discovered, they looked a little like stars.

Unlike stars, quasars are not constantly active and they can switch on and off depending on how much food they have to chow down on. Even the puniest quasar must consume the matter equivalent of 10 Suns every year and the most active can chomp their way through more than 1,000 solar masses of material in a year. At that rate of consumption, it is only a matter of time before they run out of gas and settle down to become a normal galaxy.

But, while they are active, supermassive-black-hole-powered quasars can have a dramatic effect on their corner of the Universe. The huge quantities of energy pumped out through their jets (equivalent to a trillion trillion trillion watts) travels for hundreds of thousands (or millions) of light years before smashing into the interstellar medium and spreading out (like the splash-back of a hose hitting a wall) as plumes of X-rays and radio-emitting 'dumb-bells'.

Close to the galactic centre, all that energy serves to heat up space and prevent new stars from forming, but – further out – the jets ionize the interstellar gases, inflating colossal bubbles in the interstellar medium. In turn, these bubbles create sound waves that spread through space (yes, you could say that black holes 'sing' but the sound is so deep that it would be an inaudible rumble). Like all sound waves, the black hole's rumblings compress parts of the medium they are travelling though and, where they are squeezed, the interstellar gases can get the impetus they need to collapse into new stars.

So, in galaxies with active black holes, you get almost no star formation at the centre, but at the edges, conditions are ideal for the formation of lots of massive blue stars. It is in this way that supermassive black holes perform their role as cosmic gardeners. They 'weed' some regions by retarding star formation and they fertilize others by promoting the growth of new stars.

Then there is the direct relationship between the mass of a galaxy and the supermassive black hole at its heart, which is always precisely 1,000 to 1. Such a well-refined ratio suggests that the development of a galaxy is intimately connected to its black hole – a sort of symbiotic relationship that suggests that, without supermassive black holes, the evolved galaxies that inhabit the cosmos today might never have developed.

Add to the mix the idea that the earliest radiation-spewing black holes may have accelerated the process of reionization that ended the cosmic dark ages and we find ourselves with a Universe that might owe a helluva lot to these much-maligned light-devouring beasts.

THE GOOD GARDENER

So that brings us to the supermassive black hole that inhabits our very own galaxy, the Milky Way (see pages 156–57). Our local supermassive black hole might be a long way off (about 25,000 light years) but it has had a profound effect on the chain of events that led to you and me.

As you read this, our local supermassive black hole is dormant, but billions of years ago it too was sucking up gases, stripping stars, chewing them up, and burping radiation into space.

While our cosmic gardener was active, it weeded our little patch of the galaxy and prevented the growth of the sort of massive hot stars that live short, explosive lives and created conditions that were just right for the growth of smaller, less intense and long-lived stars. One star that took advantage of these conditions was an unremarkable yellow star that would one day be called the Sun.

CONTINUES ON PAGE 159 ➡

THERE'S MORE THAN ONE WAY TO SKIN A QUASAR

Optical telescopes like Hubble can only reveal so much. To get the full picture, astronomers need to use telescopes tuned to different parts of the electromagnetic spectrum. Just have a look at how different the same galaxy (Centaurus A) looks at different wavelengths.

1. Visible light (Very Large Telescope, ESO)

The visible light image is pretty much how we would see Centaurus A if we could travel there and have a look. You can see the bright cluster of hundreds of millions of stars at the centre but most of the galaxy's features are hidden behind a cloud of dust.

2. Ultraviolet (Galex, NASA)

Looking at the galaxy in the ultraviolet doesn't solve the dust cloud problem because – like visible light – it is rubbish at travelling though clouds of dust. But it does reveal some bright patches of UV light (blue patches, top left) that are actually newly-formed stars bursting into life.

3. Infrared (Spitzer Space Telescope, NASA)

Now we switch to infrared light (which can travel through the dust cloud) and suddenly that pesky cloud of dust disappears and the shape of galaxy within is revealed.

4. X-ray (Chandra X-ray Observatory, NASA)

Move into the X-ray part of the spectrum and the galaxy now looks very different indeed. X-rays reveal what seems to be a jet of X-rays flying out from the centre of the galaxy. X-rays are emitted by very energetic or hot objects, so something very dramatic must be going on.

| Radio | Microwave | Infrared | Visible | Ultraviolet | X-ray | Gamma ray |

5. Radio (Very Large Array, ESO)

Swing to the opposite end of the spectrum into radio waves and that jet becomes very dramatic indeed. Two vast plumes of energy are seen being pumped into space. It is thought that the jets are super-hot streams of matter whose atoms have been torn apart by a monstrous black hole 55 million times the mass of our Sun.

The quasar is thought to be responsible for the star formation revealed in the UV images.

Below: When data gathered over different wavelengths are combined together, astronomers create stunning composite images that reveal structures the human eye couldn't hope to see.

THE MILKY WAY: THE GALAXY WE CALL HOME

The Milky Way is a barred spiral galaxy about 100,000 light years in diameter. it contains somewhere between 100 million and 400 million stars (and almost certainly many, many more planets).

Supermassive black hole with the mass of 4 million Suns.

Hypervelocity stars orbit the black hole at millions of miles per hour. These massive, young stars can be accelerated to such high speeds that they can be 'thrown' out of the Milky Way (like a rock flung from a slingshot).

Our Solar System, located about 25,000 light years from the galactic centre, orbits the Milky Way at about 500,000mph.

The galactic centre is a dense bar of stars and molecular hydrogen.

The spiral arms are regions where gas and dust is denser than the galactic average.

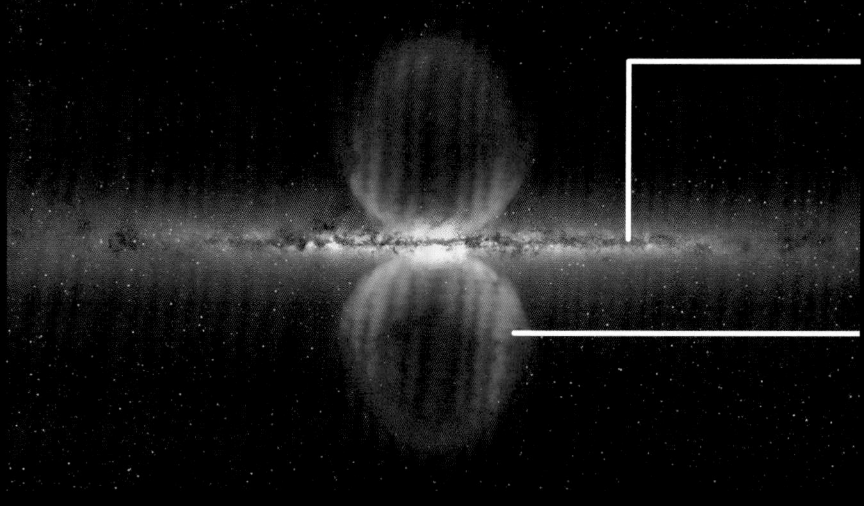

The galactic plane is, on average, about 1,000 light years deep. If you were to scale the Milky Way down to the size of a CD, it would be about three CDs thick.

Artist's impression of two huge gamma ray-emitting 'bubbles' that extend about 25,000 light years in either direction from the Milky Way. These are thought to be the result of the supermassive black hole having a 'snack' about 10 million years ago.

THE GALACTIC NEIGHBOURHOOD

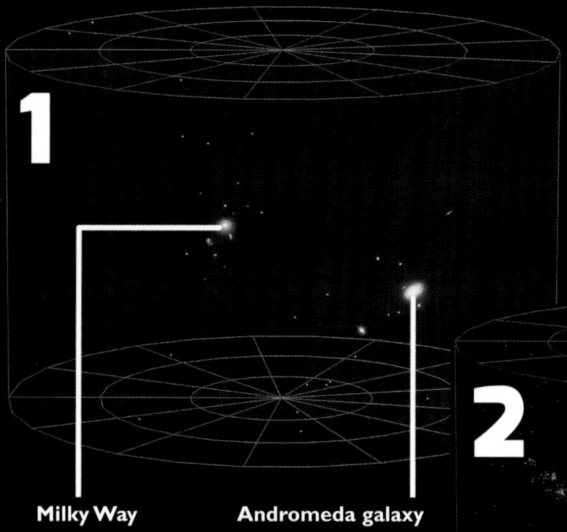

1. The Milky Way is part of a larger structure of more than 54 gravitationally bound galaxies known (imaginatively) as the 'local galactic group'.

2. This, in turn, is part of a larger structure, containing more than 100 galaxy clusters, known (more imaginatively) as the 'Virgo superstructure'.

Milky Way

Andromeda galaxy

Local galactic group

Virgo supercluster

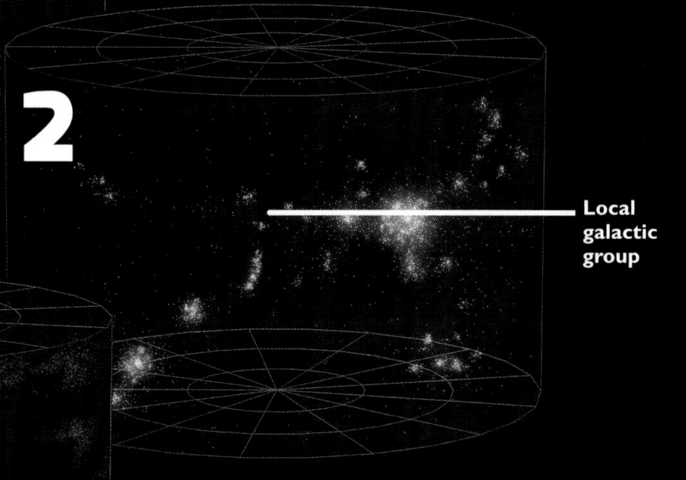

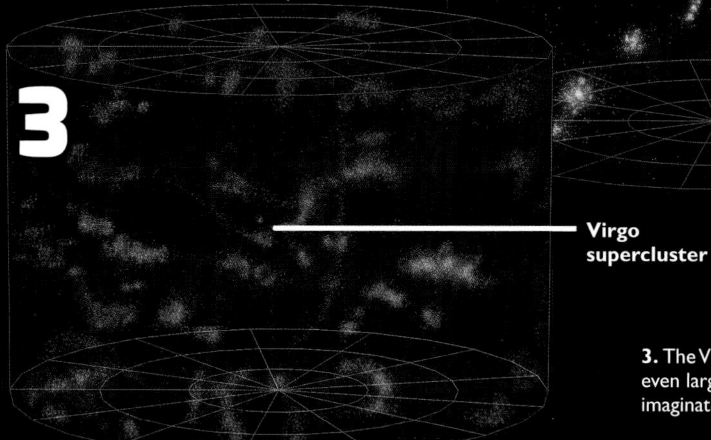

3. The Virgo superstructure is part of an even larger structure known (not at all imaginatively) as the 'local superstar'.

FINGERPRINT OF THE BIG BANG WRIT LARGE

If we zoom out even further and look at the overall structure of the galaxies beyond our own, we see something strangely familiar ...

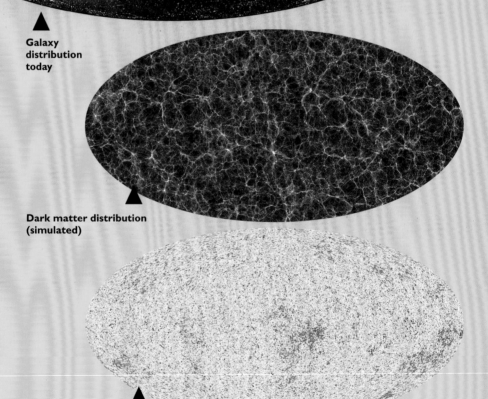

▲ **Galaxy distribution today**

The image above was taken by the Two Micron All Sky Survey (or 2MASS). It shows the distribution of more than 1.6 million galaxies as seen in the infrared part of spectrum. There are obvious concentrations of galaxy clusters, gaps where there are fewer galaxies and long filament-like ribbons connecting everything together.

If you remember that the first stars and galaxies formed on the foundations of dark matter that, in turn, formed from the energy fluctuations that emerged from the seed of the Big Bang – you can see how those tiny quantum fluctuations in that first fraction of a fraction of a second have left their mark on the Universe today.

▲ **Dark matter distribution (simulated)**

▲ **Fluctuations in the Cosmic Microwave Background)**

Now, a good gardener knows when to stop meddling and to sit back and let nature take its course. Luckily for us, our black hole horticulturist is one such gardener and, once it had laid the ground for the formation of the Sun, it stopped blazing and settled down for a nice long nap.

This period of inactivity just happened to begin about four billion years ago – the very moment that life was starting to emerge in the oceans of Earth. When it shut down, so did its radiation jets. Had the black hole not chosen this moment to hibernate, the Earth would have been showered with high-energy cosmic radiation that, at best, could have affected the chemistry of our atmosphere enough to alter, or prevent, the evolution of life; or, at worst, shredded the cells of those newly-emerged life forms and ended life on Earth there and then.

So, if the Milky Way's supermassive black hole hadn't been active at just the right time, the Sun might never have formed and, had it not entered its period of dormancy with the same fortuitous timing, life on Earth might never have evolved and (once again) you and I wouldn't exist.

But before you get all gooey-eyed about our Milky Way's benevolent black hole, it's worth remembering that these galactic beasts don't exist in a fixed state. Sometimes they will be gorging themselves and belching energy into the cosmos; sometimes they will be gently grazing and slowly accumulating mass and, at other times, they will be hibernating – quietly sleeping off a few millennia worth of food.

Our black hole might be snoozing now but, in about 5 billion years, the Andromeda galaxy will collide with the Milky Way, providing a new influx of material that will awaken the beast from its slumber and you'll want to be somewhere else when that happens.

While we are on the subject of our local star, it would seem to be a good time to turn our attentions towards building the region of space we call home – let's build a solar system.

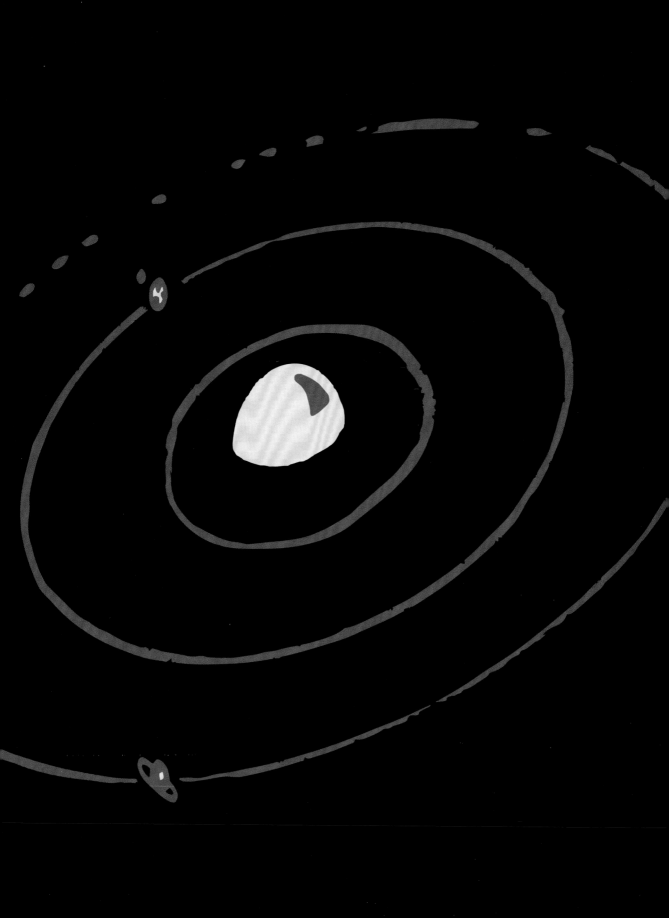

COOKING UP A SOLAR SYSTEM

In which we don our chef's whites, open the cosmic cookbook and prepare 'Solar System à la Milky Way': a spicy yellow Sun, served hot with a well-balanced combination of rocky planets and gas giants, garnished with an asteroid side salad and a delightful ring of iced primordial remnants. We will also serve our signature dish – Earth – and prepare the ingredients for life itself.

s Universe-building metaphors go, cosmic building sites and allotments are all well and good, but now it's time to get a little more intimate with our ingredients and move our activities into the cosmic kitchen.

STARTING WITH THE SOLAR SYSTEM

The secret of great chefery is preparation: if you grill your fish before your potatoes have boiled, you end up with burned fish and crunchy potatoes. The same is true when it comes to preparing a solar system – luckily we have done all the preparation we need.

INGREDIENTS

For this recipe you will need a cloud of interstellar gas that has been enriched with heavy elements. If you only have a cloud of hydrogen and helium, you will need to wait a few billion years – only when it contains traces of carbon, oxygen, silicon, iron, etc, will you be able to build planets.

The cloud will also have to be really quite big, because even a perfect star-forming nebula is incredibly diffuse – containing just a few atoms per cubic centimetre – so you'll need one that measures a few hundred billion kilometres in diameter.

KNEAD AND PROVE

Now we have to encourage the nebula cloud to collapse. We could just wait around and hope that it does so under the weight of its own gravity, but that could (literally)

Big Bang	Particles form	CMB	Dark ages (first dark matter structures)		First stars and active galaxies
13.82bn years ago		377,000 years after Big Bang			200 million years

BEHOLD, THE STELLAR KITCHEN

Below: This is the Carina Nebula – a vast interstellar cloud of star-forming dust, hydrogen, and helium gases, and other heavy elements. These nebulae are a sort of cosmic kitchen where gases can collapse to cook up stars and planets.

This one measures about 150 light years across and contains the mass (not counting already-formed stars) of about 140,000 Suns.

Galaxies evolve (clusters and superclusters form)   Solar system forms   Death of the Sun   Fate of the Universe

1 billion years   9 billion years   18.7 billion years

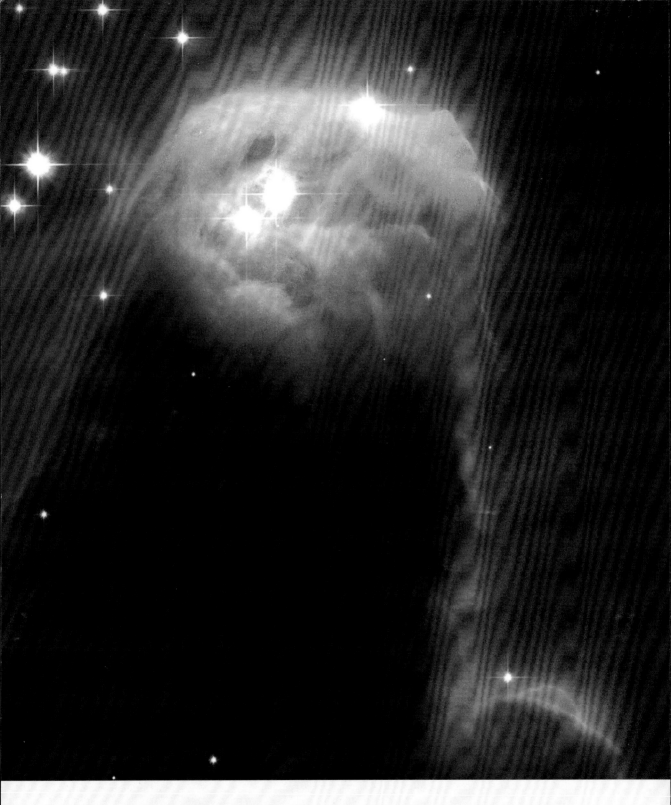

This star-forming region is called the Cone Nebula. It is just a small part of a much larger nebula (about seven light years is visible here). Hidden within the nebula's dark, opaque regions dozens of new stars are forming. The stars at the top have just (relatively speaking) broken free of the cloud.

take an eternity – if you pause to consider that, even after 13.8 billion years of cosmic evolution, much of the material in the Milky Way can still be accounted for within diffuse nebulae, it's pretty clear that our cloud won't collapse at the drop of a spatula.

No, we need to give our gases the proverbial kick up the bum. Obviously, if you want to force a mass that measures many times the width of a solar system to do anything, you need a pretty big boot, or, in our case, a pretty big star. If, as described in Chapter 6, a star were to explode as a supernova in the vicinity of a reluctant gas cloud, the resultant shockwaves that it sends through space might just be enough to get the gas cloud to start clumping and, ultimately, collapse to form a new star.

Chef's tip: It might be possible to achieve the same effect by subjecting the cloud to the high-energy outpourings of a supermassive black hole, but, be warned, black holes are powerful tools that should only be handled by the experienced cosmic chef.

So, let's assume that we've managed to encourage our nebula to collapse. You might be tempted to stir the mixture at this point, but don't – any spin already possessed by the cloud will be amplified as it collapses and, if it's spinning too fast, the nebula will be torn apart by the ferocity of its own angular momentum.

Even a gently rotating cloud will begin to spin really quite quickly as it collapses. But this spin is a good thing, because it encourages the cloud to flatten out into a disk and form a star at its centre.

The process is a little like poaching an egg – if you just boil a pan of water and then drop the contents of an egg into it, the egg will just spread out in the water (and you'll end up with a foaming white mess); but if you stir the water up, the egg will fall to the centre of the vortex (and you'll have a delightfully perfect ball of eggy yumminess).

This is where the quality of the ingredients comes into play. If you try this recipe with a cloud made of just hydrogen and helium, you'll end up with a massively massive and massively hot star that will explode long before you could ever hope to build any rocky planets around it (which, without ingredients like carbon and iron, you wouldn't be able to do anyway). Assuming your mixture *is* enriched with the right elements, it should take just a hundred thousand years or so (you might want to have a cup of tea while you are waiting) for a nice stable star, that is neither too massive nor too hot, to form.

Why do enriched nebulae make smaller stars? The key to star formation is nebula collapse and the key to nebula collapse is cooling – you have to cool the ingredients before you can bake them. As long as the atoms within a gas cloud are warm (relative to their surroundings) they possess enough thermal energy to resist the inward tug of gravity. A cold (and therefore slow-moving) particle 'feels' the effects of gravity much more than a hot (and fast-moving) particle.

As we've already seen, heavy elements and molecular compounds are much more effective heat radiators than simple atoms – so the more heavy elements a cloud contains, the more heat is radiated and the faster it cools.

In the hot early Universe, where the heaviest material was molecular hydrogen, stars could only form from clouds with enough mass to overcome thermal resistance with their sheer gravitational bulk – so we wound up with extremely massive, extremely hot and short-lived stars.

In a Universe enriched with heavy elements, you don't need lots of mass to overwhelm thermal pressure because everything is much cooler – so you end up with smaller, cooler and long-lived stars.

BAKE AND CHILL

If you did everything right, you will have a newly-formed yellow star, which we'll call the Sun (although if you are an ancient Roman, you might call it *Sol*, or if you are an ancient Greek, *Helios* might suit you better). Whatever you call the star, it should be surrounded by a flat, rotating disk of gas, called a protoplanetary disk. We will use this disk to make all the rocky planets, hot gas giants, icy gas giants, moons, asteroids and comets that will eventually populate our final Solar System.

Now, you might be wondering how on Earth-yet-to-be you could be expected to make such a disparate and varied selection of objects from just one disk of gaseous and molecular material. Well, just as biscuits, flan-bases and fish batter are made from the same basic ingredients, it all depends on how those ingredients are prepared – or, to be more accurate, *where* they are prepared.

The more observant reader will have noticed the recurring theme of this book: small, simple stuff accumulates and coalesces to make big, complex stuff – particles join to become atoms; atoms join to become heavier atoms and molecules, etc. Building planets is no different: we start with particles of gas (small stuff) and stick lots of them together to create planets (complex stuff).

The vast menu of Solar System goodies that we can cook up from just one batch of ingredients is a direct result of how close, or distant, they are from the stellar oven at the heart of it.

CONTINUES ON PAGE 170 ➡

FROM NEBULA TO SOLAR SYSTEM

The cloud of dust and gas that became our Solar System started to collapse about 4.6 billion years ago. It took just 100,000 years for the Sun to form and another 10 million for the gas giants, like Jupiter, to form. The rocky planets formed after 100 million years had passed.

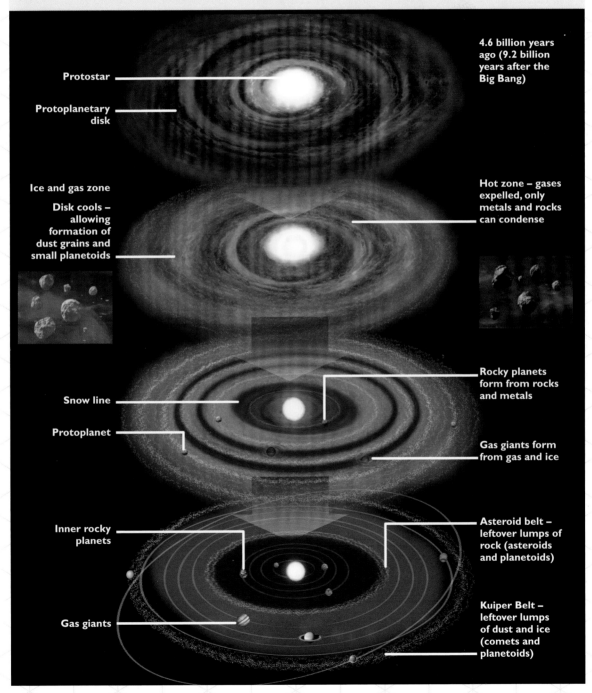

4.6 billion years ago (9.2 billion years after the Big Bang)

Protostar

Protoplanetary disk

Ice and gas zone

Hot zone – gases expelled, only metals and rocks can condense

Disk cools – allowing formation of dust grains and small planetoids

Snow line

Rocky planets form from rocks and metals

Protoplanet

Gas giants form from gas and ice

Inner rocky planets

Asteroid belt – leftover lumps of rock (asteroids and planetoids)

Gas giants

Kuiper Belt – leftover lumps of dust and ice (comets and planetoids)

THE SUN: OUR AVERAGE YELLOW STAR

We have feared it, worshipped it, and argued about whether or not it orbits the Earth (it doesn't, by the way). The Sun has nurtured and threatened life on Earth in almost equal measure. We owe our very existence to our long-lived and (relatively) stable star, so it would be rude not to get to know it a little better ...

1. CORE

15 million°C

The core is the engine room of the Sun. The extreme temperatures and pressures at the core are sufficient to sustain nuclear fusion (but you know that already).

The Sun converts 4 million tonnes of hydrogen into energy every second and has been doing so for 5 billion years. It will continue burning through fuel at this rate until it runs out in about another 5 billion years.

2. RADIATION ZONE

2–7 million°C

Energy from the core travels through the radiation zone in the form of electromagnetic radiation.

The region is so dense that it takes an average 170,000 years for energy from the core to leave the radiation zone, but it can take millions of years to escape (it's exactly the same problem the earliest photons had in the super-dense, super-hot post-Big Bang Universe).

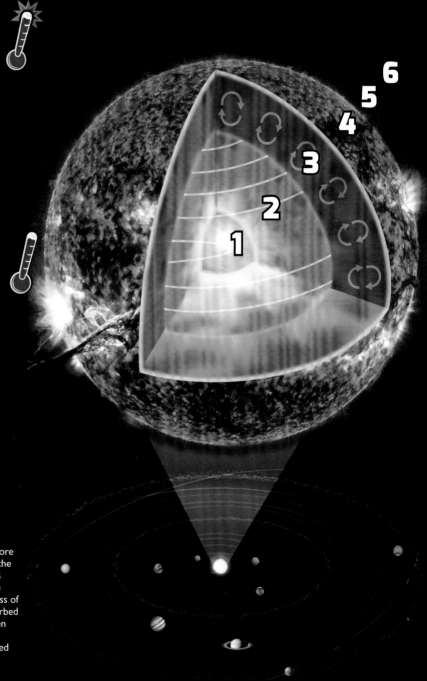

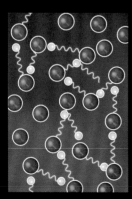

Photons travelling from the core must face the 'drunkard's walk' – the long process of being absorbed by hydrogen nuclei and then emitted in random directions.

3. CONVECTION ZONE
About 2 million°C

This turbulent region carries energy to the Sun's surface in thermal columns. The material cools at the surface and plunges back to bottom of the convection zone. It is reheated by the radiation zone where it travels back to the surface once more.

4. PHOTOSPHERE
5,700°C

The Sun's visible surface.

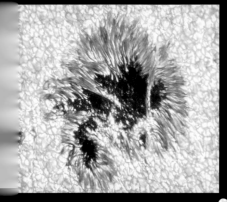

A VIOLENT STAR

Earth to scale

Coronal mass ejections (CMEs) are the most powerful events in the solar system. A single CME event can throw more than 10 billion tonnes of charged particles (mostly protons and electrons) into space – covering an area as wide as 30 million miles.

Shockwaves can accelerate these particles to close to the speed of light – at this speed they can cover the 150 million miles to Earth in as little as 90 minutes – so being caught in the path of a CME would be like standing in the path of a colossal particle accelerator.

5. ATMOSPHERE
3,700–98,000°C

At about 500 kilometres above the photosphere, the lower atmosphere is the coolest region of the Sun. Above this is the chromosphere. About 2,000 kilometres thick, this region of the Sun's atmosphere increases in temperature with altitude reaching about 100,000°C.

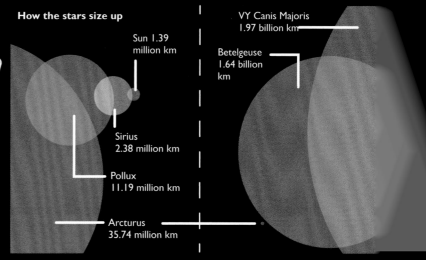

How the stars size up

Sun 1.39 million km

Sirius 2.38 million km

Pollux 11.19 million km

Arcturus 35.74 million km

VY Canis Majoris 1.97 billion km

Betelgeuse 1.64 billion km

6. CORONA
1–10 million°C

The corona is the Sun's extended atmosphere – larger in volume than the entire Sun. Temperatures rise with altitude to in some places, 10 million°C

How the solar atmosphere can be so much hotter than the Sun's surface is one of science's biggest mysteries. It is now thought that so-called Alfvén waves – waves in the plasma that carry energy via the Sun's magnetic field – might be responsible for the rise.

Those regions of gas closest to the oven (on the stellar hob, so to speak) receive the most radiation, or heat, from the Sun. The further the ingredients are from the Sun, the colder the conditions in which stuff can be cooked (or frozen) up.

ON THE HOB

In the regions closest to the Sun, it is too hot for water molecules to form (from atoms of oxygen and hydrogen) and, thanks to a vigorous stream of high-energy particles streaming from the Sun, called the solar wind, most of the atomic hydrogen and helium and lighter elements have been blown into the further regions of the disk. This leaves behind only the heavier elements, so it is here that we will cook up the small rocky planets, including the Earth.

IN THE CHILLER

Just beyond the wind-blasted and baked inner regions is a region called the snow line. Here we find all the gases and lighter elements pushed out from the inner regions and, of course, the heavier elements. It is also cold enough for hydrogen and oxygen atoms to bond and condense to form water (in the form of ice).

With such a vast supply of gases at our disposal, it is here that the gas giant planets are made.

IN THE FREEZER

The further we move from the Sun, the colder the disk becomes. Beyond the gas giants is where the ice giants, like Uranus and Neptune, are made. But the disk also becomes less dense as we approach its extremities, so it is here that we find the smaller icy bodies, such as comets and the demoted ex-planet, Pluto.

COOKING UP A ROCKY PLANET CALLED EARTH

Every aspiring cosmic chef wants to jump straight in and make a nice rocky planet – after all, we live on a rocky planet and, as far as we know, it is on a rocky planet that life has the best chance of evolving. Luckily, this is also where your planet-baking training will begin.

So, how do you turn a vast swirling disk of gases and chemical elements into a planet?

START SMALL (VERY, VERY SMALL)

If you were able to borrow a giant planet-smashing hammer (or good ol' Darth Vader's Death Star doohickey) and use it on a rocky planet, you would find yourself with a pile of large rocks. If you smash one of those rocks up, you would get some smaller rocks and then some even smaller rocks and gravel. Take one of those small rocks and grind it up and you will end up with a pile of dust. Grind it even more, and you would end up with a tiny pile of chemical elements and compounds. Keep grinding and you'll get atoms, then subatomic particles ... but you will have gone too far.

If you had filmed this process and then played the film backwards, you would see dust forming from atoms, rocks from dust, large rocks from smaller rocks and, finally, from those larger rocks, a planet would emerge: that's how you create a planet.

Just as gravity made the galaxies and stars by bringing together disparate elements through their mutual attraction, once again, it looks as if gravity is the fundamental force that steals the show and makes planet-building possible. For the most part that's true, but in the crucial dust-forming stages of planet building it is actually the electromagnetic force that gets the ball rolling (so to speak).

In the swirling mess of atoms that make up the protoplanetary disk, no individual atom or molecule has enough gravitational superiority to attract other atoms. It's like a street-corner evangelist trying to attract followers by whispering at a chattering crowd – he just won't be heard. But, if he were to draw followers in one at a time (perhaps

CONTINUES ON PAGE 174 ➡

GASES TO GASES, DUST TO DUST

So how did disparate collections of atoms become the planets?

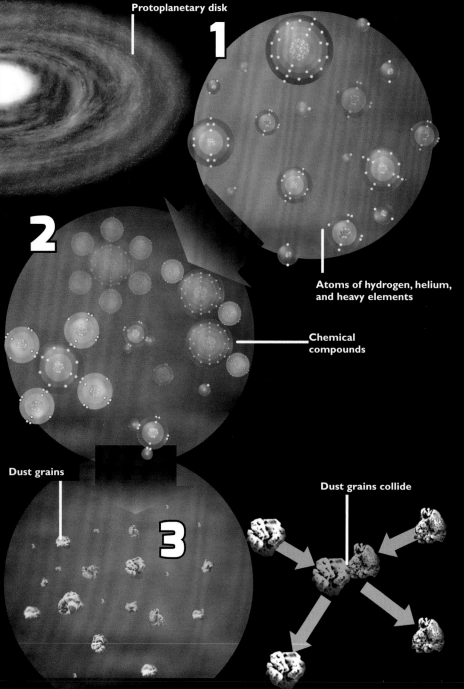

Protoplanetary disk

1

Atoms of hydrogen, helium, and heavy elements

2

Chemical compounds

Dust grains

Dust grains collide

3

1. So, here are all those heavy elements we cooked up in cores of stars.

2. Considering how much energy was required to make the chemical elements, once made they are quite keen to bond together. Elements with a deficit of electrons in their outer shell want to balance the figures, so they bond with those that have a surplus (and vice versa). By sharing electrons, two elements become a chemical compound (iron bonds with oxygen to become iron oxide, or rust).

3. The molecules of chemical compounds accumulate into microscopic grains of dust.

4. This is where things get a bit tricky. When these grains collide in the vacuum of space, instead of 'wanting' to stick together, they prefer to bounce off each other (like pool balls).

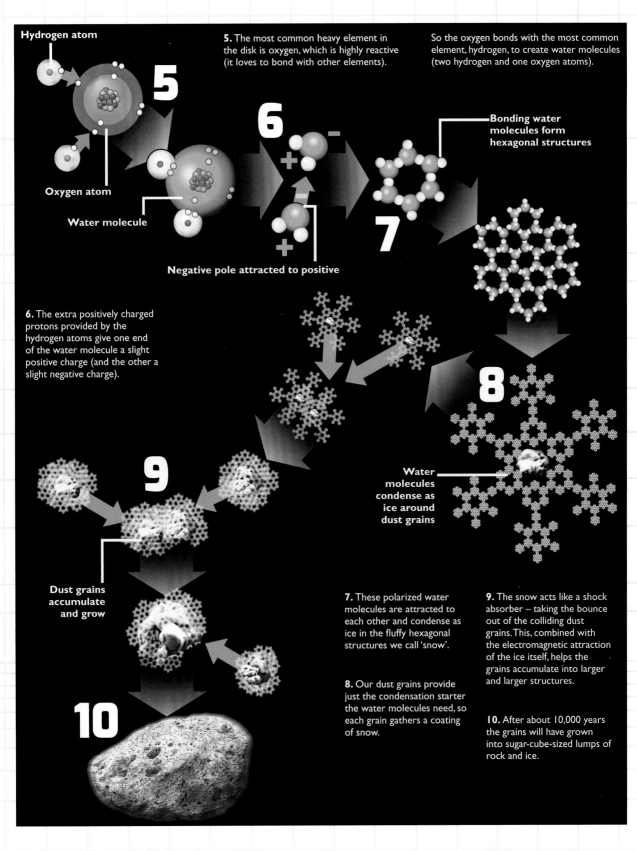

Hydrogen atom

5

5. The most common heavy element in the disk is oxygen, which is highly reactive (it loves to bond with other elements).

So the oxygen bonds with the most common element, hydrogen, to create water molecules (two hydrogen and one oxygen atoms).

Oxygen atom

Water molecule

6

+
−
−
+

Negative pole attracted to positive

Bonding water molecules form hexagonal structures

7

6. The extra positively charged protons provided by the hydrogen atoms give one end of the water molecule a slight positive charge (and the other a slight negative charge).

8

Water molecules condense as ice around dust grains

9

Dust grains accumulate and grow

7. These polarized water molecules are attracted to each other and condense as ice in the fluffy hexagonal structures we call 'snow'.

8. Our dust grains provide just the condensation starter the water molecules need, so each grain gathers a coating of snow.

9. The snow acts like a shock absorber – taking the bounce out of the colliding dust grains. This, combined with the electromagnetic attraction of the ice itself, helps the grains accumulate into larger and larger structures.

10. After about 10,000 years the grains will have grown into sugar-cube-sized lumps of rock and ice.

10

by tapping them on the shoulder and whispering in their ear) and getting them to pass the message to others – eventually, he might attract the entire crowd.

This is how the electromagnetic force draws together enough atoms in the protoplanetary cloud to start making grains of dust. In the right conditions, atoms and molecules are quite eager to stick together. Some might have one too many electrons in their outer shells, whereas others have one too few. Since atoms crave balance and stability, those with electron deficits, or surpluses, will want to balance things out by bonding and sharing their outermost electrons – hydrogen atoms bond so well with oxygen (to create water) because oxygen has two electron pairs in its outermost shell and two single electrons; it really wants four pairs so the

ROCK AND ROLL: BUILDING A ROCKY PLANET

If you've ever tried to make a larger rock by squeezing two smaller ones together (and haven't we all?), you'll appreciate that it isn't something they do too readily. It won't come as much of a surprise, then, that it takes about 1,000 times longer to build an Earth-sized planet than it does to build a Sun-sized star.

1. Small centimetre-sized lumps of rock and ice meet and stick together.

2. At first they collate to form loosely-bound clumps, but as they accumulate mass they are squeezed into increasingly solid lumps.

Planetesimals

1 2 3

deficit pairs hook up with hydrogen atoms to share their single electron. Or, to put it another way: the electromagnetic force gets atoms with a slight negative electrical charge (like oxygen) to 'stick' to atoms with a slight positive charge (like hydrogen).

So, after a bit of electromagnetic bonding, we find ourselves with some dust – hardly an impressive claim when you consider we just finished cooking up a galaxy, but, nevertheless, a crucial first step towards making a planet. It's even less impressive when you consider that those bits of dust are no bigger than the tiny ash particulates you might find in cigarette smoke. But what *is* impressive is this: when you add enough of them together, you get a planet ... that's an entire world from grains of dust!

3. By the time they reach about a kilometre in diameter, they possess enough mass to start attracting each other through gravity alone – at this point, we can start calling them planetesimals.

4. As their gravitational pull increases, so does the size of the objects they attract and so does the energy with which they collide. Sometimes these collisions will be so energetic that the planetesimals are smashed into smaller chunks, but eventually they will become big enough to absorb all but the most powerful impacts.

5. All that collision energy creates a great deal of friction and, as such, a great deal of heat. By the time the lump has grown to protoplanet status, its interior will have begun to melt – add to this the heat created by gravitational energy as its mass is crushed under the weight of its growing bulk – and, by the time it has reached fully fledged planet size, our baby planet will have a dynamic molten interior.

Protoplanet

Planet

4

5

MAGNETIC ATTRACTION

The next job, of course, is actually sticking all those itsy-bitsy dusty pieces together and that's not as easy as you'd think. Firstly, they are still too small to have any significant gravitational influence on their surroundings, which means they can't add to their bulk by pulling matter from the disk. Instead, they have to rely on random collisions as they fly around their infant star – and here lies the second problem.

When tiny dust grains collide in the vacuum of space, they much prefer to bounce off and ricochet away than stick together. Once again, it is the electromagnetic force* that comes to the rescue. (*Technically, here we are talking about the electrostatic force – the attraction (or repulsion) felt by objects of different (or the same) electrical charges.)

By far the most plentiful element in the protoplanetary disk is hydrogen and the most common metal (astronomers refer to anything heavier than helium as a 'metal') is oxygen. Given their propensity for bonding, this means that there is likely to be a huge amount of water floating around with the dust grains – not as liquid water, but as water molecules, or vapour.

When these free-floating water molecules come into contact with our micron-sized dust grains, they condense on to its surface but, in the near-vacuum of space, they do so as ice. And, because the water molecules have a positive electrical charge at one end and a negative charge on the other, they link together to form intricate hexagonal structures – in other words, they condense as snowflakes.

With all those electrically charged water molecules neatly lined up, the dust grain becomes electrically polarized and behaves like a tiny bar magnet. So, when it makes contact with another polarized grain, their respective negative and positive poles align and the grains stick together. The snowflakes have another useful property: they are fluffy. This means that the dust grains are covered in a fluffy outer layer that behaves like a shock absorber, so, when the grains collide, the snow cushions the impact and stops the grains from 'bouncing' away from each other.

In this way, over the course of about 10,000 years, the micron-size dust grain can accumulate to form rocky sugar-cube-sized lumps, or, as I like to call them, planet seeds.

There is an obvious problem with this method of course – it only works beyond the disk's snow line, where the hydrogen and oxygen needed to make water can exist and condense. So what does that mean for the rocky planets that we want to cook up on the 'hot side' of the disk?

There are three possible solutions to this conundrum. The first is that the planet seeds formed beyond the snow line and then, as they gained mass, they migrated inwards. The second is that the above process took place within the hot zone but with silicate grains (compounds of silicon and oxygen, the basic component of all rocky planets like Earth) fulfilling the impact-absorbing role. A third more radical alternative suggests that the planets formed simultaneously alongside the Sun in pockets of regional gas collapse. In this scenario, the rocky planets can form in situ before the Sun gets the chance to heat the protoplanetary disk and limit their growth – only when they are pretty much as we see them today did the blast of radiation that heralded stellar ignition blow the un-accreted gases into the disk's outer regions.

Either way, once they gain enough mass, the planet seeds will be able to attract each other gravitationally – eventually building up to become huge planetary embryos.

But this won't happen overnight. To gather together enough material to form gravitationally influential kilometre-sized rocks will take about 100,000 years and you'll be waiting anywhere between 10 million and 100 million years for a decent-sized rocky planet. This might seem like a long time but, when you consider that we went from a disparate collection of tiny particles of dust weighing just a fraction of a gram to an Earth-size planet weighing in at almost six million billion billion kilograms, it's remarkable that the process happens so quickly (even so, you might want to invest in a decent book while you wait).

DECORATE YOUR PLANET

Now that we have a basic rocky planet you will probably want to decorate it with a nice atmosphere and a few oceans. But before you start, you need to consider the following:

If the planet is too close to the Sun, the atmosphere will be stripped from the planet's surface by powerful, radiation-laden solar winds – like paint is removed by a sandblaster. A planet that is too close to the solar furnace will have its surface baked to thousands of degrees during the day – while, on the night side, the lack of insulating atmosphere will mean that heat is haemorrhaged into space and surface temperatures will plunge well below freezing.

If the planet is too small, it won't have enough gravity to hold on to its atmosphere and, if it isn't sandblasted away, it will just drift off into the vacuum of space.

But how do you get a gaseous atmosphere and watery ocean on to a planet that formed in a region of the disk where gases and water have been expelled? Well, just as the tiny seeds from which the rocky planets grew could have migrated inwards from colder regions of the disk, so can an awful lot of other stuff. By the time your planet is ready to receive an atmosphere, there will be lumps of dusty ice (comets) and ice-laden rock (asteroids) that never managed to reach planet-building stature flying around all over the place.

Some might be falling inwards because they lost energy in collisions, others will have been thrown around by the gravity of larger objects – either way, in the first few hundred million years of the solar system's life there will be a lot of icy stuff smashing into our baby planet.

The water ice and frozen gases they carry with them will be deposited on to the planet's surface as oceans of liquid water and the primordial atmosphere will be enriched with oxygen.

HOW TO ENSURE A MELT-IN-THE-MIDDLE CORE

Everyone loves a planet with a warm gooey centre: a planet with a dynamic molten core will generate magnetic fields that shield it from the worst of the Sun's radiation hissy fits and make dynamic geological processes, like plate tectonics, possible. Luckily, the formation of a molten core is a natural by-product of the violent nature of a planet's formation – the trick is making sure your planet is big enough to retain its internal heat.

Let's rewind the clock a few tens of millions of years and return our planet to its component rocky lumps. Now, as we start smashing the lumps together, as well as increasing the size of our planetary seed, you will see that the impacts are creating a lot of heat. This is because, as they zoom around space, the rocks possess huge amounts of kinetic energy. When they collide, the kinetic energy is converted into heat and the larger the object is, and the more energetic the collision, the more heat is released.

SIZE MATTERS

There is evidence that Mars once had a molten core but, because it is only half the size of Earth, there wasn't enough 'insulation' around the core to prevent it radiating most of its heat into space.

HOW THE EARTH GOT ITS CORE

The Earth's metallic core is a heat generator and magnetic dynamo that powers and protects our home planet. But how did a lump of hot rock get its dynamic metallic heart in the first place? Once again, we have gravity to thank for that...

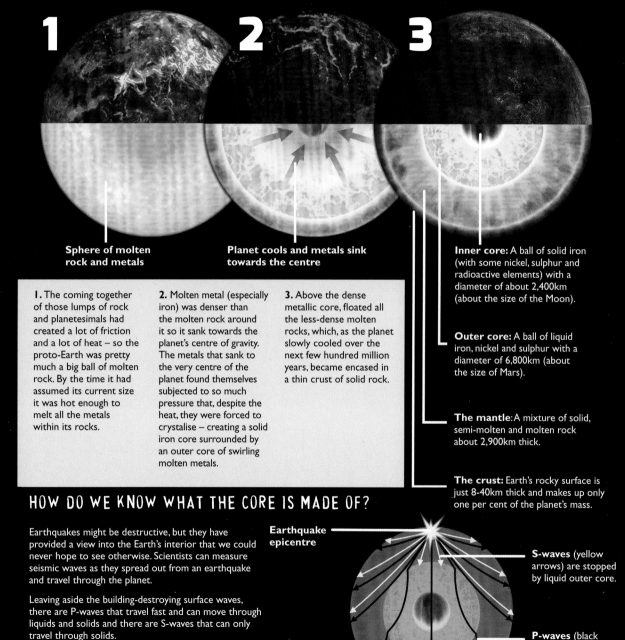

1

Sphere of molten rock and metals

2

Planet cools and metals sink towards the centre

3

1. The coming together of those lumps of rock and planetesimals had created a lot of friction and a lot of heat – so the proto-Earth was pretty much a big ball of molten rock. By the time it had assumed its current size it was hot enough to melt all the metals within its rocks.

2. Molten metal (especially iron) was denser than the molten rock around it so it sank towards the planet's centre of gravity. The metals that sank to the very centre of the planet found themselves subjected to so much pressure that, despite the heat, they were forced to crystalise – creating a solid iron core surrounded by an outer core of swirling molten metals.

3. Above the dense metallic core, floated all the less-dense molten rocks, which, as the planet slowly cooled over the next few hundred million years, became encased in a thin crust of solid rock.

Inner core: A ball of solid iron (with some nickel, sulphur and radioactive elements) with a diameter of about 2,400km (about the size of the Moon).

Outer core: A ball of liquid iron, nickel and sulphur with a diameter of 6,800km (about the size of Mars).

The mantle: A mixture of solid, semi-molten and molten rock about 2,900km thick.

The crust: Earth's rocky surface is just 8-40km thick and makes up only one per cent of the planet's mass.

HOW DO WE KNOW WHAT THE CORE IS MADE OF?

Earthquakes might be destructive, but they have provided a view into the Earth's interior that we could never hope to see otherwise. Scientists can measure seismic waves as they spread out from an earthquake and travel through the planet.

Leaving aside the building-destroying surface waves, there are P-waves that travel fast and can move through liquids and solids and there are S-waves that can only travel through solids.

By seeing which waves arrive where and at what time, scientists can figure out what sort of material they have travelled through.

Earthquake epicentre

S-waves (yellow arrows) are stopped by liquid outer core.

P-waves (black arrows) travel through liquid core but change speed and are diffracted.

By the time it has become an Earth-sized world, the planet has collected so much energy that it has become, in effect, a giant sphere of molten rock. Within this ball of magma, temperatures are high enough to melt the metals, including iron, locked away within the rocks. Because metallic atoms are heavier than the rock that contained them, they sink towards the planet's centre of gravity, which also happens to be the centre of the planet and, over millions of years, this settles down to become a core of molten iron (with a little nickel and sulphur thrown in). At the very heart of the planet, there is so much material weighing down on it that the extreme pressure forces the iron back into a solid state – creating an inner core of solid iron.

HOW TO KEEP A CORE NICE AND HOT

Keeping a planet piping hot is always a problem – after all, with all that cold empty space surrounding it, it will rapidly cool and become a cold, dead lump of solid rock and metals. So how do you keep a planet like Earth from 'going cold'?

Do you remember those really heavy radioactive elements that were made in the explosive death throes of massive stars? Well it's time to put them to use.

When radioactive elements like uranium decay (by shedding alpha particles to become lighter elements) they release energy and, even though they make up just the tiniest fraction of all the elements in the Universe, our growing Earth is able to scoop up enough radioactive stuff to provide it with a long-lasting source of energy – a nuclear power plant that keeps the core nice and hot.

The molten iron outer core swirls around the solid inner and acts like a giant dynamo, generating electric currents. As the electric currents rotate around the core, they create a powerful magnetic field that flows from the planet's poles.

Eventually, the molten rock surrounding the core starts to cool around its outer edge, forming a thin crust of solid rock (a bit like the skin that forms on a bowl of cooling rice pudding). It is on this tenuous strip of solidified rock that the oceans will collect and, one day, life will evolve. The magnetic field generated by the Earth's molten dynamo will protect the planet from life-destroying solar radiation and keep its atmosphere from being lost to the void.

An unfortunate side-effect of all this geological activity is that our young Earth will be covered by unsightly

CONTINUES ON PAGE 184 ➡

THE DYNAMO EFFECT

Without its protective magnetic field, the Earth would be a barren Mars-like desert. Luckily, a molten core makes a perfect magnetic field generator.

1

Outer core boundary 4,000°C

Inner core boundary 7,000°C

Convection currents in outer core

2

Currents twisted into 'rollers'

3

Earth's rotation

Magnetic field lines

1. The outer core might be a churning cauldron of molten metal but, compared to the inner core, it is relatively cool. This means that material nearest to the inner core is hotter than material closest to the mantle. This creates convection currents in which hot material rises from the depths of the inner core, cools, and then sinks back down again.

2. Because the Earth is spinning, that means all the fluids within it spin also. The rotational forces (known as coriolis forces) twist the currents into 'rollers' of fluid that are aligned with the Earth's axis.

3. Electrical currents flow through the circulating pillars and act like the coil of wire in an electric dynamo – creating a dipole magnetic field (like a series of super-strength bar magnets).

4. The magnetic field extends into space and creates a magnetic bubble, called the magnetosphere, which deflects and diverts the worst of the high-energy radiation that flows from the Sun in the solar wind.

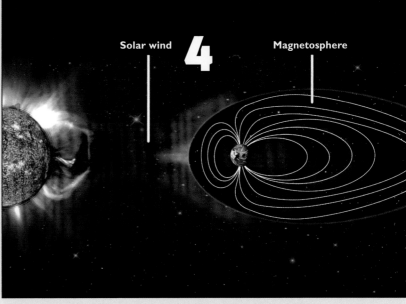

Solar wind

4

Magnetosphere

It is thought that early in its history, Mars had an atmosphere not unlike Earth's, but, because the planet was too small to maintain its magnetic field, it was lost to the sand-blasting effects of the solar wind.

PLANETARY DROPOUTS

Not all planetesimals advance to protoplanet status, some just run out of material to accumulate. Some of these 'almost-planets' are lured into orbits around their more advanced cousins – where they become moons. Others are doomed to wander space as asteroids (like the asteroid belt-object, Vesta), or minor planets (like Ceres – also in the asteroid belt between Mars and Jupiter.

This 'minor planet' is scarred by a belt of five-kilometre-deep troughs that wrap around it surface. It is thought these scars are features known as graben – caused when two surface faults move apart.

Graben can only be formed when an object has a layered molten interior, which means that, given the chance, Vesta would have become a rocky planet like Mars or even Earth.

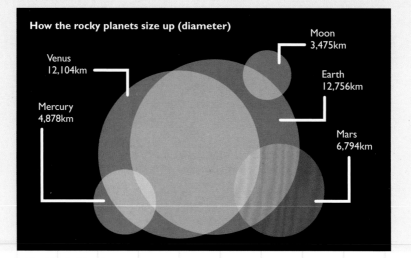

How the rocky planets size up (diameter)

Moon
3,475km

Venus
12,104km

Earth
12,756km

Mercury
4,878km

Mars
6,794km

HOW TO MAKE A MOON

Although many of the planets acquired their moons by capturing asteroids and planetoids, or through the accretion of orbiting debris, the Earth gained its Moon through a rather more traumatic process.

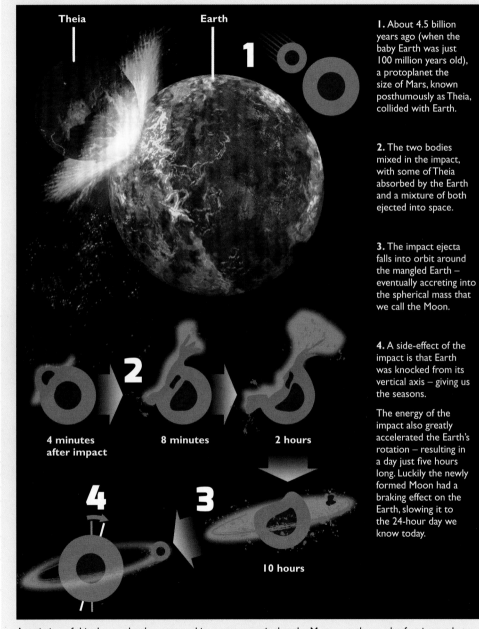

Theia **Earth**

1

2 4 minutes after impact 8 minutes 2 hours

4 **3** 10 hours

1. About 4.5 billion years ago (when the baby Earth was just 100 million years old), a protoplanet the size of Mars, known posthumously as Theia, collided with Earth.

2. The two bodies mixed in the impact, with some of Theia absorbed by the Earth and a mixture of both ejected into space.

3. The impact ejecta falls into orbit around the mangled Earth – eventually accreting into the spherical mass that we call the Moon.

4. A side-effect of the impact is that Earth was knocked from its vertical axis – giving us the seasons.

The energy of the impact also greatly accelerated the Earth's rotation – resulting in a day just five hours long. Luckily the newly formed Moon had a braking effect on the Earth, slowing it to the 24-hour day we know today.

A variation of this theory that has emerged in recent years is that the Moon was the result of an impact between two more evenly matched planetary bodies each about five times the size of Mars.

Whichever variation you prefer, what isn't in doubt is the fact that the birth of our Moon was a pretty violent event.

volcanoes (not unlike the face of an acne-riddled teen), but fear not, these will calm down a bit after a few billion years (although they won't clear up entirely).

The volcanoes vent huge quantities of volatile gases (and some water vapour liberated from rocks), which will build up to create a carbon-dioxide-rich atmosphere – if you want lots of oxygen in your atmosphere you will need to populate the oceans with photosynthesizing bacterial life, which absorbs carbon dioxide from the atmosphere and releases oxygen in water from its hydrogen bonds, but you won't be able to do this until you actually have some oceans – this will be possible after about 200 million years when the planet has cooled down a bit more. We'll come back to the Earth when it's cooled down and start seeding the life that, one day, will lead to you.

COOKING UP A JUPITER

Making a giant gas planet can be a pretty daunting prospect – after all, a Jupiter-sized planet has about 1,300 times the volume of Earth. But really, it's not so very different to making a rocky world – it's just a case of where you prepare it.

The first stages of preparation are no different to those we probably used to make Earth, where the planet seeds formed beyond the snow line. But, instead of allowing the protoplanet to drift into the warm, rocky-planet-forming zone, we simply keep it orbiting beyond the snow line where gases like hydrogen and helium are plentiful.

As the protoplanet moves through its orbit, its gravity draws the gases in towards it and, as it spins on its axis, it wraps the gases around itself – like a stick twirled in a vat of cotton candy. Eventually the planet will sweep up all the gases that it can reach – clearing a furrow through the planet-forming disk but, by this time, its rocky origins will be buried beneath 70,000 kilometres of gas.

There is so much gas (about 90 per cent hydrogen and 10 per cent helium) piled up around the core that most of it is subjected to immense pressure, which makes it take on some very un-gas-like properties. Around the core, hydrogen gas is squeezed and condensed into a metallic liquid soup – a swirling electrically charged ocean more than 40,000 kilometres deep – which, coupled with the planet's rapid spin, creates a magnetic field almost 20,000 times stronger than the Earth's.

GAS GIANT COMPOSITIONS

When it comes to gas giants' formation, it really is a first-come-first-served sort of affair. Jupiter, being the first to form, took the lion's share of the hydrogen and helium gas, and the rest was mopped up by Saturn. Uranus and Neptune, which formed later, had to make do with the leftovers ...

Jupiter (gas giant)

— Gaseous hydrogen, helium, ammonia and ice

— Liquid hydrogen

— Metallic hydrogen

Saturn (gas giant)

— Gaseous hydrogen, helium, ammonia and ice

— Liquid hydrogen

— Metallic hydrogen

Uranus (ice giant)

— Gaseous hydrogen, helium and methane

— Water, ammonia and methane ices

Neptune (ice giant)

— Gaseous hydrogen, helium and methane

— Water, ammonia and methane ices

How the gas giants size up

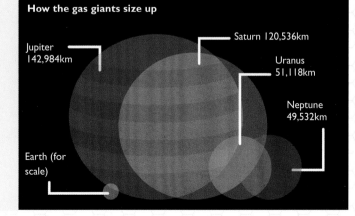

Jupiter 142,984km

Saturn 120,536km

Uranus 51,118km

Neptune 49,532km

Earth (for scale)

BUILDING THE GAS GIANTS

They might dwarf the rocky planets but gas giants are much quicker to make than their terrestrial cousins. All you need is a rocky planetoid of a reasonable size; lots (and lots) of hydrogen and helium gas, and about 10 million years …

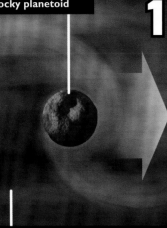

Rocky planetoid

1

Gases in protoplanetary disk
(mostly hydrogen and helium)

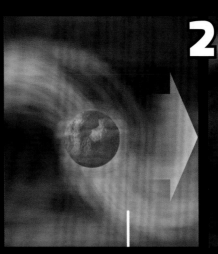

2

Gases build up around core

Gases accreted
around rocky core

Clears disk material
within its orbit

1. By the time a space rock has accumulated enough mass to be called a planetoid, it has quite a lot of gravitational influence on its surroundings – especially when its surroundings are made up of gas.

2. As the planetoid travels around its orbit, it draws in the gases that surround it – wrapping itself in a hydrogen cocoon.

3. When it has become more gas than rock, the planetoid finds itself demoted to role of 'core' (greed is the third deadly sin after all).

4. The more mass the infant gas giant accumulates, the faster it sucks up gases and dust and, within a just a few million years, it runs out of material to accrete and stops growing.

5. Gas might not be very heavy, but the planet has gathered so much that it has a great deal of mass. Jupiter is so massive that only in its outermost layer of hydrogen can maintain its gaseous state.

6. Below this, the increasing pressure squeezes the hydrogen atoms together – compressing the gas into its liquid state.

7. Buried beneath 30,000 kilometres of liquid hydrogen, the pressure is so extreme that the space is squeezed from the atoms, so the electron orbits are almost pressed up against the proton nuclei. So tightly are they squeezed that the electrons get confused about which nucleus they should be orbiting and they start wandering between atoms – creating a steady stream of electrons that flow as electric currents. Since the ability to conduct electricity is usually reserved for metals, hydrogen in this state is known as 'metallic' hydrogen.

8. As these electric currents swirl around in the rapidly rotating planet, they generate a magnetic field.

9. The magnetic field created by a gas giant the size of Jupiter is some 20,000 times stronger than that created by the Earth and is projected so far into space that it overlaps the orbit of Saturn, 100 million kilometres away.

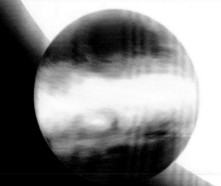

Gas giants are common throughout the Milky Way.

One of the most massive, HAT-P-2b, lives about 370 light years away and contains the mass of eight Jupiters (that's 2,500 times more massive than Earth).

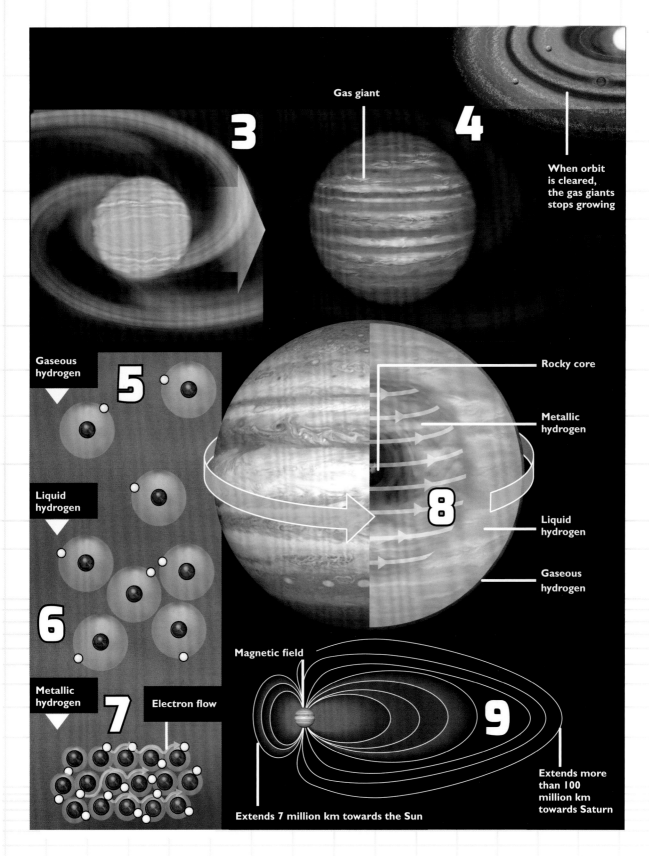

3

Gas giant

4

When orbit is cleared, the gas giants stops growing

Gaseous hydrogen

5

Liquid hydrogen

6

Metallic hydrogen

7

Electron flow

Rocky core

Metallic hydrogen

8

Liquid hydrogen

Gaseous hydrogen

Magnetic field

9

Extends more than 100 million km towards Saturn

Extends 7 million km towards the Sun

As the pressure increases with depth, so does the temperature. At the edge of the metallic hydrogen layer it can reach about 10,000°C, while, deeper still at the boundary of the core, it might be as warm as 36,000°C (that's about seven times hotter than the surface of the Sun).

ALTERNATIVE RECIPE

The mysteries surrounding planet-formation are legion, so it's worth mentioning an alternative theoretical method for making gas giants. There are some cosmic chefs (or astrophysicists as they prefer to be known) who believe that, rather than forming through a bottom-up process (starting with a small rocky core and accreting gas), the gas giants may have formed through a top-down process called the gravitational instability model.

This is because, by some calculations, it would take too long for a gas giant the size of Jupiter to be built from the bottom up. Instead, they suggest that they formed directly from regions of the gaseous protoplanetary disk that collapsed under the weight of their own gravity – rather like miniaturized versions of star formation.

This method has the benefit of allowing a gas giant the size of Jupiter to form from scratch in about the same amount of time it takes to build a centimetre-cubed lump of protoplanet using the other method. On the downside, it doesn't seem to fit with observations about the composition of gas giants and, arguably worse, if a planet the size of Jupiter formed too early in the life of the planet-forming disk, there is a very real chance it would have just spiralled into the infant Sun.

Whichever Jupiter-forming method you chose, one thing is certain – it uses up so much of the disk's gas reserves that there is very little left to build the rest of the gas giants. Cook up a Saturn-size planet shortly afterwards and you will have used up almost the entire supply.

If you want to make any more planets you will have to use the only remaining ingredients – mostly ice and some leftover bits of hydrogen, helium and heavier elements. With these 'freezer leftovers' you should be able to put together a few ice giants, like Neptune and Uranus (snigger).

THE LEFTOVERS

Even after most of the planet-forming disk has been eaten up by the planets, there is plenty of stuff left over at the end of the meal. Like the family forced to eat cold turkey for a week after a large Christmas dinner, the solar system puts its leftovers to good use. Some of the larger rocky bodies might be captured by the planets and bent to their gravitational will as supplicant moons, while others might wander the solar system as asteroids in huge looping orbits. Further afield, the icy bodies that didn't make the grade as ice giants might also wander the expanses of interplanetary space as comets, or form part of the ring of rock and ice that makes up the Solar-System-encircling ring of formation debris known as the Kuiper Belt.

The most unfortunate leftovers might find themselves banished to the furthest reaches of the Sun's gravitational influence – becoming part of the large, but diffuse, cloud of icy bodies that is known as the Oort Cloud.

COOKING UP LIFE

Normally the last thing a chef wants to see is a swarm of tiny little beasties taking up home on one of his creations, but in the cosmic kitchen, life is arguably the pinnacle of planetary chefery: only the very finest planetary creations are good enough for life to want to make it their home (well, from a slightly arrogant anthropogenic point of view).

How life first emerged on Earth is one of the most contentious and poorly understood of all the great cosmic mysteries. The easiest way to explain it would be to summon one of mankind's many deities and say that 'He', 'She', or 'They' simply snapped their supernatural fingers and summoned life from the ether on an arbitrary day of the Universe creation cycle. But if we were to adopt such a shortcut, we might as well throw this book in the trash and replace every chapter with 'stuff exists because "He" (or "She", or "They") wished it to be so'.

So, how did life organize itself and emerge from the chaos of the newly formed Earth? The short answer is: we don't know; and the long answer would require an entire book of its own – so we'll have to make do with an unsatisfactorily abbreviated version.

THE END OF THE SUN ...

When the Sun starts to run out of hydrogen fuel in about 5 billion years, it will slowly expand by 259 times to become a red giant star. It will swallow the inner rocky planets and, in about 7.5 billion years from now, the Earth will also be incinerated.

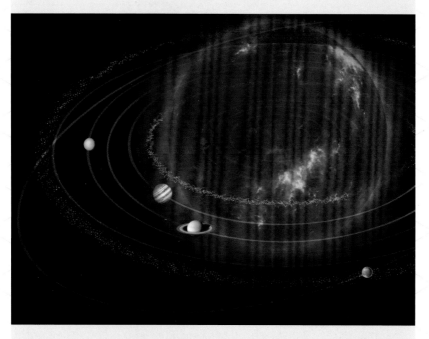

If it's any consolation, life will have ceased to exist on the Earth long before it is swallowed by the Sun. Over the next billion years, our local star will become increasingly hot and bright.

The increased solar radiation will evaporate the oceans and the Earth will become a scorched, barren desert. Eventually, the heat will become so intense that the Earth will become a molten sphere of rock once more.

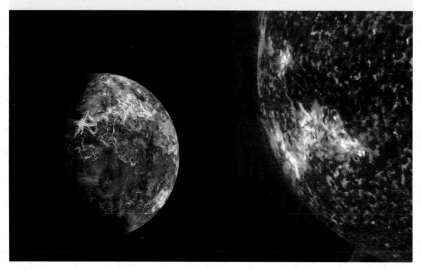

LIFE IS STAR STUFF

How the building blocks of life were created is no mystery – they were created in exactly the same way as everything else in this book was: raw materials created in the Big Bang are enriched within stars and assembled through interactions of the fundamental forces.

Do you remember those complex chemical compounds that formed within the protoplanetary disk – oxygen bonding with hydrogen to create water molecules and iron combining with oxygen to create iron oxides? Well, the chemical bonding process didn't end with the likes of rust and water – the ingredients for life were made in that cloud too.

If you strip life down to its most basic components you will find that it is made up of a collection of chemical molecules such as proteins and nucleic acids. Proteins are made up of amino acids and these, in turn, are made up almost entirely of hydrogen, oxygen, carbon, and nitrogen – the four simplest elements to be cooked up in the stars (apart from non-reactive helium, which stubbornly refuses to bond with anything). The nucleic acids, which include DNA (the double-helix chemical code book that tells cells how to work) and its cousin RNA are, at their most basic level, made of nucleic acids called nucleotides. Each nucleotide is made up of three simple elements – a carbon 'backbone' with oxygen and hydrogen attached to it. Bonded to one part of this sugar is a phosphate group, which is just a phosphorus atom with oxygen attached to it, and bonded to another part is a nitrogenous base, which is made up of nitrogen, oxygen and hydrogen.

Next to oxygen, carbon is the second most common heavy element to have been cooked up by stars – this is because 95 per cent of all the stars in the Universe don't have enough mass to reach the carbon burning phase (resulting in a lot of unburned carbon being liberated when the star dies). Also, like oxygen, carbon is an enthusiastic bonder – it loves to bond with other elements. Carbon also has the benefit of being able to form long chains in which carbon atoms link together to create a sort of chemical spine, to which other atoms or chains of atoms can bond – providing the chemical foundations for the construction of complex organic molecules.

Using spectroscopy, astronomers have detected organic compounds (the building blocks of the building blocks of life) in abundance floating around within star-forming nebulae. In fact, hundreds of different life-essential molecules have been detected in such regions – amino acids (the building blocks of proteins) are, it seems, relatively common.

So, even before the planets started to form, there already existed all the chemical components we need to make life. Much of it would have been swept up during the planet-forming process and assimilated into the dust and rocks that would make up the Earth but, crucially, not all of it was locked away in planetary vaults. Some was left behind in the Solar System's planet-building leftovers – carried within the frozen bodies of comets, meteors and asteroids – and, long after the Earth had passed its molten inferno phase, these raw materials of life were deposited into the planet's oceans.

Even today hundreds of tonnes of organic material are delivered to Earth every year by comet and asteroid dust. In the early years of the Earth's history, these impacts would have been even more common so, in the hundred million years or so it took for life to 'pop up' in the Earth's oceans, billions of tonnes of life's chemical building blocks would have been delivered.

How life went from a soup of ingredients to the first single-celled, self-replicating organisms is still a mystery and subject to intense – and impassioned – debate. One of the most attractively simple ideas is that life emerged from warm tidal pools massaged by the gravitational pull of the Moon.

ADD WATER AND SIMMER GENTLY

During the period that we know life emerged, the Moon was much closer to the Earth than today (it has been moving away ever since its formation) so tidal influence would have been much greater, and the early Earth may have experienced tides 1,000 times higher than today – meaning they could sweep much further inland.

Also, about four billion years ago, the Earth was spinning much faster. A day was only about six hours long – meaning that the colossal floods inundated and withdrew from the land every three hours. As the tides withdrew, they would have left behind pools of water containing the organic molecules.

But, even back then, the oceans were really quite big, so all our billions of tonnes of organic material would have been extremely diluted – making it highly unlikely that the molecules could have bonded to create more complex materials. This is where those tidal pools come in.

Water deposited in those shallow pools would have been evaporated by the Sun – leaving behind the salts, minerals, and organic compounds. Thanks to the frequent tides, the pools would have been topped up with water every few hours, which would have also evaporated. After many top-ups and subsequent heating and evaporation, the pool of water would have become a highly concentrated, warm chemical soup.

Energy provided by solar radiation, lightning and the action of tides would have encouraged chemical reactions – creating ever more complex organic compounds such as fatty acids (chains of carbon, oxygen and hydrogen atoms) and proteins. Some of this primordial soup might have leached into microscopic cracks in the clays and rocks that lined the pools, where water-repelling fat molecules, or lipids (made up of fatty acids), would have formed bubbles, which trapped organic compounds. Inside these fatty bubbles, or cells, chemical reactions continued – providing the energy to create new molecules and allowing it to grow and multiply. The cell could be said to be metabolizing (using chemical reactions to create energy), growing and reproducing – the basic properties of life.

The longer the chemical reactions continue, the more complex the compounds become – leading all the way up to DNA and, eventually, to complex life-forms like fish, amphibians, reptiles, mammals, and of course you and me.

Obviously this is a gross over-simplification of a poorly-understood process, but I include it to show that the idea that life could have spontaneously emerged from a pile of star debris isn't as far-fetched as it might sound.

So there we have it: we have built particles from energy; we have manipulated the fabric of the Universe and used it to build galaxies and stars; we have used those stars to build the chemical elements; and we have used those elements to construct the planets and the life that inhabits them. Our journey is almost complete.

But before we down tools and put the kettle on, let's spend a moment to contemplate where our creation is heading – what fate awaits our Universe?

THE END... OR IS IT?

In which we flirt with another of the Universe's dark sides, ponder the death of everything, and slice the singular from the cosmos and create a multiverse.

o we find ourselves gazing in wonderment at the Universe we have built together (assuming, of course, you built it right and you're not gazing at an empty page floating in a void of shoddily-constructed space-time). From nothing, we have forged everything, and it is truly awesome ... It seems a shame then to ponder its destruction – yet ponder it we must, so let us turn our gaze to the far future and consider the death of the Universe.

WHAT GOES UP, MUST COME DOWN

Not so very long ago, astronomers thought they had the fate of the Universe pretty well nailed down. The cosmos had exploded into existence almost 14 billion years ago and its contents had been thrown forth in the Big Bang like shrapnel from a bomb blast.

As with a bomb blast, after the initial rapidly inflating blast wave of energy, there follows a period when, driven by momentum, matter and energy travel outwards as an expanding sphere.

Another parallel the Big Bang has with small bangs is that, as the initial blast of energy expands, it dissipates and cools. In the moments following the Big Bang, the temperature of the Universe was around 100,000 billion billion billion°C but, after 13.8 billion years of expansion, its temperature today is a less than toasty -270.4°C.

But momentum can only take you so far and eventually it must run out. When it does, the shrapnel slows, stops, gravity takes hold and everything falls back to Earth – it's a classic case of 'what goes up, must come down'.

It was a logical assumption then that the Universe will follow a similar path and, once the initial momentum of the Big Bang runs out, the stars and galaxies will cease

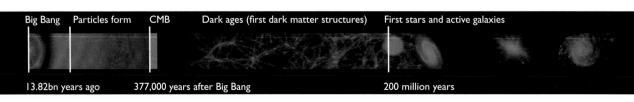

Big Bang | Particles form | CMB | Dark ages (first dark matter structures) | First stars and active galaxies

13.82bn years ago | 377,000 years after Big Bang | 200 million years to move apart and, with nothing to counter the draw of gravity, everything will 'fall' together – until, at some distant point in the future, the Universe will end in a sort of Big Bang in reverse or, as it was dubbed, a Big Crunch. Some even speculated that, with all the matter and energy of the Universe returned to its infinite point of density state, it might explode outwards once more – creating a new Universe in some sort of Big Bounce.

It was logical. It was neat. It was even slightly comforting to think that the Universe might be forever reborn in an infinite cycle of Big Bangs, Crunches and Bounces. Unfortunately, it was also completely wrong.

WHAT GOES UP, KEEPS GOING UP

In the 1990s, astronomers were less concerned with the Universe's distant fate, which didn't promise to reveal anything particularly new, than they were with mapping the cosmos as it is today. To this end, two teams of researchers set themselves the task of charting the locations of the most distant supernovae.

They were looking for a special sort of supernova, known as type-1a. These cosmic explosions can be traced back to binary star systems in which at least one of the stars is a white dwarf (the dense core remnant of a Sun-like star). The white dwarf strips its companion, which might have reached its red giant phase, of stellar material and, in doing so, gains so much mass that it becomes unstable and goes kaboom. Type-1a supernovae have the advantage of being a predictable brightness so that, like the Cepheid stars used by the early astronomers, they can be used as yardsticks to measure the Universe; for this reason, they are also known as standard candles.

The teams found 50 such supernovae but, when they measured their 'predictable' brightness, they found they were not as bright as they should have been. They were not where they would be expected to be if their movement away from

CONTINUES ON PAGE 200 ➡

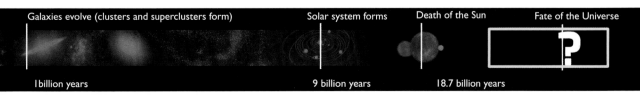

Galaxies evolve (clusters and superclusters form)   Solar system forms   Death of the Sun   Fate of the Universe

1 billion years   9 billion years   18.7 billion years

THE END... OR IS IT?

GALAXIES GALORE

This is the Hubble eXtreme Deep Field (XDF). Almost everything you can see in the image is an individual galaxy (each home to hundreds of millions of stars) and there are more than 5,500 galaxies visible.

You would be forgiven for thinking that this must be some sort of all sky view – a sweeping panorama of the heavens.

In fact, it is very, very much smaller.

The amount of sky covered in this image is equivalent to area of sky obscured by a grain of sand held at arm's length on the tip of your finger.

Or, to put it in more astronomical terms, a portion of the sky about a 14th the diameter of the Moon as you see it in the night sky ... that's an awful lot of stuff in a teeny tiny patch of sky.

A portion of sky this big (relative to the Moon) contains all those galaxies

The most distant galaxies visible in the Hubble XDF are 13.2 billion years old – right at the edge of the observable Universe.

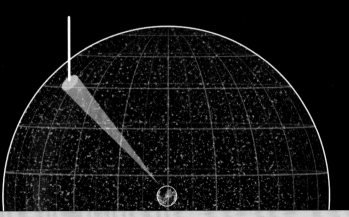

us was slowing, or, to put it another way, they were just too far away. This could only mean one thing: instead of running out of momentum and slowing down, the expansion of the Universe was accelerating.

The paradigm-overturning results were revealed to the world in 1998 and, since then, further measurements have not only confirmed that the Universe's expansion is accelerating, but that the rate of acceleration is accelerating: rather than ending in a Big Crunch, it seems the Universe will just keep expanding and it will do so at an ever-increasing rate.

It's hard to overstate just how shocking this realization was – it didn't just overturn conventional wisdom, it slapped it round the face, kicked it up the bum and sent it home wearing a pink tutu and scuba mask. Imagine you threw a ball into the air and, as expected, its ascent slowed but then, rather than falling back to your hand as common sense and the laws of physics dictate, it started to move upwards again – moving faster and faster until it accelerated out into space – leaving you with a daft expression on your face and a hole in your ceiling. You'd probably be more than a little surprised ... now replace the word 'ball' in that scenario with 'entire Universe' and you might come close to how the astronomical community must have felt.

Of course, once they had recovered from the shock, they then had to figure out what was powering the expansion. It was clear that something we couldn't see or directly detect was pushing the Universe apart when gravity should have been pulling it all back together: something was trumping gravity as the dominant force in the Universe.

Like dark matter in the 1930s, the mysterious anti-gravity force was given a holding name – to give folk something to call it while astronomers struggled with the details – they called it dark energy. But what was it?

REDEEMING EINSTEIN'S 'BIGGEST BLUNDER'

When Einstein applied his General Relativity equations to the structure of the Universe, something unexpected fell out of his calculations – they showed that gravity should be pulling the galaxies back together and the Universe should be contracting. But, since prevailing wisdom dictated that the Universe was static and eternal, this disturbed Einstein. So he added an extra equation to his calculations – one that counteracted the force of gravity on cosmic scales and kept everything nice and stable. He called his piece of mathematical fine-tuning his cosmological constant.

But then Edwin Hubble threw the cat amongst the cosmological pigeons and revealed that, far from being unchanging, the Universe was actually expanding. The exasperated Einstein threw out his cosmological constant, declared it to be his biggest blunder and it was all but forgotten about for seven decades.

Then, when the spectre of dark energy reared its ugly (but invisible) head in 1998, astronomers realized that Einstein's constant was just the anti-gravity pill the doctor ordered. So it was dusted off and reinstated to the galaxy-repelling role for which it had been invented: Einstein had predicted the need for dark energy more than 70 years before anyone even suspected it was needed ... but then he had thrown it away.

Put simply, dark energy is the price the Universe pays for having 'empty' space. As we've seen, empty space is never truly empty – any given volume of space contains an intrinsic energy. This was all well and good in the first few billion years after the Big Bang because the Universe possessed more 'stuff' than it did empty space so gravity was the dominant force. But as the fabric of space-time expanded, so did the space between the 'stuff' and the amount of dark energy increased along with it.

Eventually, when the Universe was about seven to eight billion years old, there was so much empty space that the balance was tipped away from gravity. Dark energy became the dominant force and the expansion of the Universe, which had been slowly grinding to a halt, started to regain its momentum – this time driven not by the Big Bang but by the ever-increasing influence of dark energy.

The Universe's gravity anchors had been lifted and there was nothing to hold the expansion back. To make matters worse, once it took over, dark energy had a runaway effect, with more empty space breeding more dark energy, which breeds more empty space, which breeds more dark energy (and so on) – once dark-energy-fuelled expansion started, the Universe was doomed to ever-accelerating expansion.

THE BIG FREEZE

As the expansion of space-time accelerates, the stars and galaxies carried along with it will move further and further apart. Within perhaps five billion years, galaxies will be receding from their neighbours so quickly that the expansion of space between them will outpace the light emitted by their stars.

CONTINUES ON PAGE 206 ➡

HOW BIG IS THE OBSERVABLE UNIVERSE?

The 'observable' Universe is that portion of the cosmos that we can see from Earth. It only includes objects whose light has had enough time to complete the journey to Earth ... so how big is it?

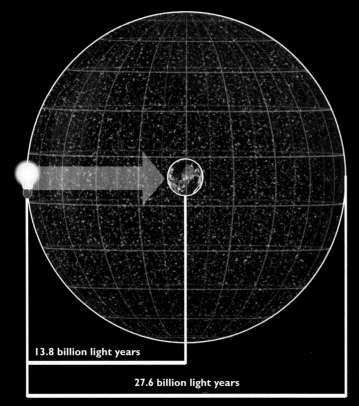

13.8 billion light years

27.6 billion light years

On the face of it, this seems like a straightforward, even pointless, question to ask.

The Universe is 13.8 billion years old and nothing can travel faster than the speed of light.

So the most distant object we could possibly see is one whose light has had enough time to reach us.

It stands to reason then that the 'observable Universe' can't be more than 13.8 billion light years in any direction.

Therefore the observable Universe's radius is 13.8 billion light years – making a sphere with a diameter of 27.6 billion light years.

That would be the logical answer, but, as we've discovered many times in this book, the Universe is not a very logical place and the observable Universe is actually an awful lot bigger ...

INFLATING DISTANCE

1. Light leaves an object 13.8 billion years ago and starts its journey towards the Earth (which won't exist for another 9 billion years).

2. It takes 13.8 billion light years to reach us at the speed of light.

3. But, ever since the object emitted that light, it has been moving away from us as the Universe expands.

4. So, by the time the light arrives at Earth, although the light has only travelled 13.8 billion light years, the object that emitted the light is now 48 billion light years away – making the full extent of the visible Universe about 93 billion light years in diameter.

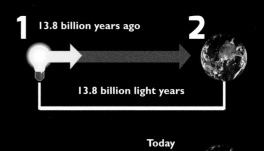

1 13.8 billion years ago

2

13.8 billion light years

Today

4

3

48 billion light years

13.8 billion light years

So, the stuff whose light has taken 13.8 billion light years to reach us is no longer 13.8 billion light years away.

As the Universe expanded so everything sitting on its space-time surface was carried away from everything else – after 13.8 billion years of expansion, the object that emitted its light when it was 13.8 billion light years away has moved off to a distance of about 48 billion light years – making the full extent of the visible Universe about 93 billion light years in diameter.

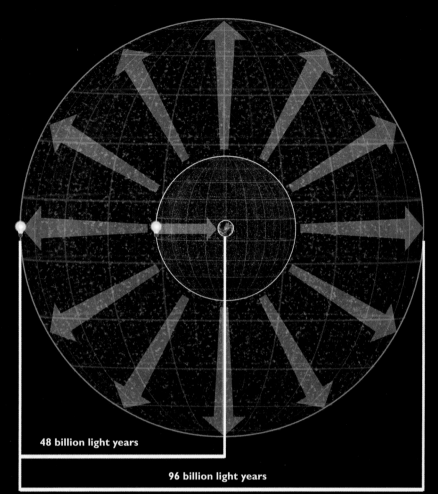

48 billion light years

96 billion light years

If the observable Universe was scaled down to the size of the Earth, our blue planet would be about 180 times smaller than a single atom.

... AND WHAT ABOUT THE REST OF IT?

We know that the observable Universe represents just the tiniest fraction of the Universe as a whole, but to appreciate the scale of the entire is probably beyond the scope of the human mind.

Warning: Astronomy cliché coming up

The observable Universe contains about 100 billion galaxies and each of those contains about 100 billion stars – meaning that just the observable portion of the Universe contains about 10,000 million million million stars – that's more stars than there are grains of sand on the whole of the Earth.

As for the rest of the Universe, we can really only speculate about its size – after all, if you can't see it, you can't measure it ... but one thing we can say for certain is that it's big ... really, really, really big.

GRAVITY'S NEMESIS: DARK ENERGY

In today's Universe, almost 70 per cent of the total energy distribution is made up of dark energy. Dark energy behaves as an anti-gravity force – accelerating the expansion of the Universe.

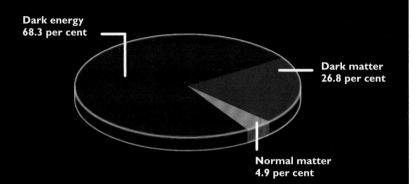

**Dark energy
68.3 per cent**

**Dark matter
26.8 per cent**

**Normal matter
4.9 per cent**

It is thought that dark energy is actually Einstein's cosmological constant – an anti-gravity energy that emerges spontaneously from the quantum vacuum as space-time expands but whose density remains constant.

Imagine you had a slice of bread (space-time) that is spread with butter (dark energy) – as you stretch the bread, instead of getting thinner, the butter actually remains spread with the same thickness. You have more butter but its density within a certain area hasn't changed.

HOW DARK ENERGY TIPPED THE SCALES

Matter (dark and normal) is distributed across the 'surface' of the fabric of space-time – so as the Universe expands, it becomes spread out and diluted. The Universe gets bigger but the amount of normal matter stays the same.

But dark energy is distributed evenly within the fabric of space-time (within both space and time) and, as space expands, the amount of dark energy expands along with it. Its density remains constant – so the more 'empty' space there is, the more dark energy there is.

Dark energy

Matter

1. When the Universe was still quite small, everything was very densely packed and there wasn't much 'empty space' – so there wasn't much dark energy.

2. But as the Universe expanded, more empty space was created between the pockets of matter – but matter still 'outweighed' dark energy so gravity was still the dominant 'large-scale' force.

3. About seven billion years after the Big Bang, so much empty space had opened up between the pockets of matter that dark energy started to 'outweigh' the amount of matter.

When that happened, dark energy's anti-gravity effect overwhelmed gravity – accelerating the expansion of space.

The scientific consensus is that dark energy does exist and that its density is constant – but it's fun to contemplate the alternatives ...

No dark energy

Gravity slows and then reverses the Universe's expansion – eventually causing it to collapse in on itself in the Big Crunch.

Dark energy constant

Dark energy takes over and Universe expansion continues. Overall percentage of dark energy increases but density is constant – resulting in a steady increase in expansion.

Matter slowly cools until the Universe experiences heat death, or the Big Chill.

Dark energy increases

Dark energy takes over and Universe expansion continues. But, instead of remaining constant, its density increases – resulting in exponential expansion. Space-time expands so rapidly that even the space within atoms expands. Matter is torn apart at an atomic level in the Big Rip.

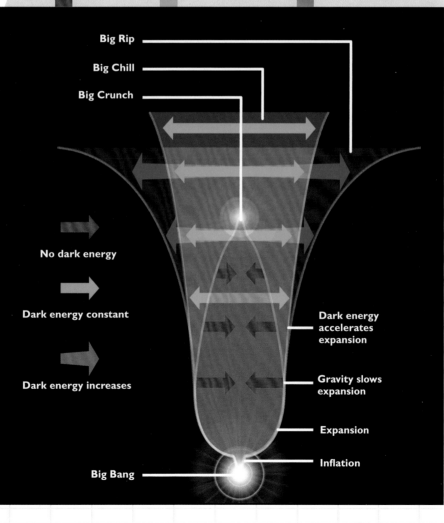

Big Rip

Big Chill

Big Crunch

No dark energy

Dark energy constant

Dark energy increases

Dark energy accelerates expansion

Gravity slows expansion

Expansion

Inflation

Big Bang

From Earth's point of view (assuming we are still here), the galaxies that have told us so much about our place in the Universe will fade from the night sky and vanish completely. Any future civilization studying the heavens will only see the stars and galaxies that are gravitationally bound to the same region as the Milky Way: they will assume (as we once did) that our local galaxy is the full extent of the Universe. With the cosmic microwave background lost and no evidence of receding galaxies, they will conclude that the Universe is unchanging and eternal.

Of course, they will be very wrong.

As the Universe expands, the energy it contains will become spread increasingly thinly and space will become colder and colder (but not dark energy of course). Eventually, over the course of perhaps thousands of billions or even trillions of years, its temperature will drop to absolute zero (-273.15°C).

If you remember that temperature is a measurement of the movement of atoms, when the cosmic thermometer hits absolute zero (which is as cold as cold can get), the atoms (and their subatomic constituents) will cease to move and matter will reach a state of thermal death.

Having finally achieved maximum entropy (remember entropy is the desire for stuff in an organized system to decay into disorder), the Universe will be a near-infinite wasteland of black stars and frozen worlds. Eventually, even the atoms themselves will disintegrate and, for the first time in its history, the Universe will be truly empty.

TAKE SOLACE IN THE MULTIVERSE

For we humans, death is usually a time of great sadness in which we mourn the loss of a loved one. Often, we can seek solace in the idea that a part of us lives on in our children, or in the countless tiny ways we have influenced the people and the world around us. But how can we console ourselves about the death of our Universe? Surely the Universe *is* everything and, with its passing, all things passed with it. Nothing lives on. It is the end.

EMBRACE THE DARK SIDE

Before you get too much of a downer on dark-energy, it's worth considering that there are some physicists who believe that it was a dark-energy-like form of anti-gravity vacuum energy that was responsible for cosmic inflation – that first surge of expansion that took the post-Big-Bang Universe seed, inflated it and set it on its way to become the Universe we know today.

But what if it isn't as unique as we think it is? What if our Universe is just one strand of reality in an infinite web of existence – just a small corner of an infinite multiverse?

As counterintuitive as it sounds, there are some physicists who believe that, far from being the beginning of all things, the Big Bang was just the moment our particular Universe burst from the womb of a parent universe – the latest offspring of a much larger multiverse.

The idea that our Universe is just one of countless others might seem (at best) incredible and (at worst) delusional, but remember this: we once thought our planet was unique; then we thought our solar system was unique and, after that, we thought our galaxy was unique – is it such a stretch to imagine that our Universe isn't as unique as we like to believe?

And, if our Universe had a 'parent', it stands to reason that it might have children of its own where some part of our sphere of existence can live on.

CHILDREN OF THE BLACK HOLE

Do you remember that infinitely dense point of mass and energy where the laws of physics as we understand them didn't exist and space, time, and all the fundamental forces, were one – the primeval atom, or singularity, that contained everything we needed to build our Universe?

Well, during our Universe-building journey, we've come across something very similar – an object of infinite density that exists apart from the laws of physics and is divorced from space and time – the singularity that hides at the heart of the black hole. According to one theory, our Universe might have been born within just such a black hole, and black holes within our Universe are creating universes of their very own.

When we built our black holes, we followed their journey from star to collapsed core all the way to the achievement of supermassive black hole status. But it might be possible that, instead of stopping at the 'point of infinite density' stage of collapse, the singularity 'bounces back' and punches a hole in the fabric of space-time. From there, it begins to expand a new pocket of space-time, creating a Big Bang from which a new universe is born.

The new universe won't be a perfect clone of its parent and there might be tiny variations in the way the laws of physics develop. Gravity might be slightly stronger, in which case stars might form too quickly and become too massive for stable Sun-like stars to form; or it might be slightly weaker, in which case the stars might not form at all. The potential permutations (and subsequent universe-forming results) are infinite.

THE VACUUM THAT INFLATED A BUBBLEVERSE

Another multiverse theory comes as one of the more bizarre off-shoots of cosmic inflation: false vacuum inflation. It was thought up by the originator of cosmic inflation, Professor Alan Guth. According to Guth, our Universe began life as a false vacuum, which is a rather counterintuitive sort of non-thing that is crammed packed with repulsive, anti-gravity energy. The false vacuum's gravitational repulsion (meaning it pushes things away) was so strong that it drove the Universe's early inflation. Guth argues that as the false vacuum expanded it decayed into energy that, in turn, created the matter that makes up the Universe today.

Another quirk of a false vacuum is that, unlike gas expanding to fill a room, its energy doesn't 'thin out' as it expands – so you get the nice evenly balanced Universe we see today. Also, according to this inflationary model, although the Universe will be nice and even when viewed as a whole, there will be localized ripples of greater or lesser density (as seen in the microwave background). Around the high-density ripples, matter will become gravitationally attracted and stars and galaxies will form. But low-density ripples could see space around them collapse into a new false vacuum that will drive the inflation of a new Universe. In this way, an infinite chain of Universes could form as bubbles on bubbles within bubbles ...

These (and other multiverse theories) offer a solution to a problem that has vexed cosmologists for decades: namely, why does our Universe seem to be so perfectly tailored for the emergence of intelligent life?

Everything seems to exist in just the right proportions and is governed by just the right sort of physical laws to ensure that the evolution of life was almost inevitable.

HOW TO BUILD A MULTIVERSE

Together, we have built a splendid Universe that is perfectly tuned for the creation of stars, galaxies, planets, and even life. But, what are the chances that such a 'perfect' Universe could emerge from nothing?

What if our Universe is just one among an infinite multiverse? In a multiverse of infinite possibilities, the emergence of a 'perfect' Universe becomes an inevitability ...

CHILDREN OF THE BLACK HOLE

One theory is that our Universe was born within a black hole and that black holes within our cosmos are creating universes of their own.

1. Here's a black hole much like any other. At its heart is the hyper-concentrated, smaller-than-an-atom lump of matter we call the singularity.

2. According to standard theory, space and time become so heavily distorted at the singularity that time stops ... but what if it only stopped in our Universe?

It has been suggested that in the final collapse of the singularity, it 'bounces' back and punches a hole in the fabric of space-time.

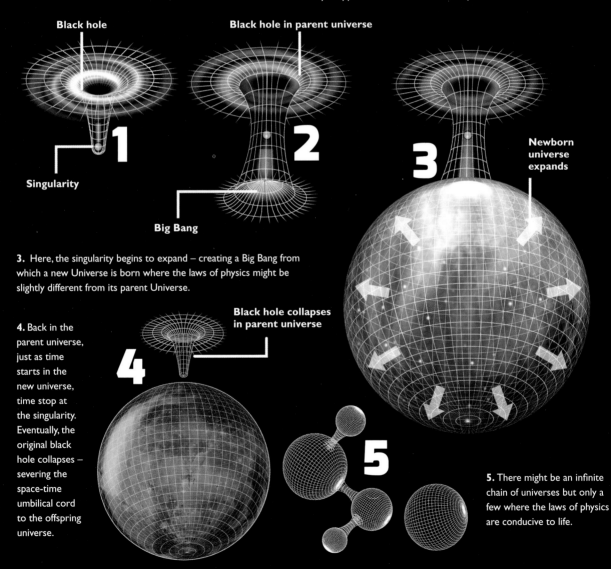

Black hole

Singularity

Black hole in parent universe

Big Bang

Newborn universe expands

Black hole collapses in parent universe

3. Here, the singularity begins to expand – creating a Big Bang from which a new Universe is born where the laws of physics might be slightly different from its parent Universe.

4. Back in the parent universe, just as time starts in the new universe, time stop at the singularity. Eventually, the original black hole collapses – severing the space-time umbilical cord to the offspring universe.

5. There might be an infinite chain of universes but only a few where the laws of physics are conducive to life.

INFLATING A BUBBLEVERSE

This method neatly ties together two concepts we've used in this book already – cosmic inflation and vacuum energy. It was thought up by Alan Guth, the physicist who came up with the concept of cosmic inflation.

We've seen how empty space, or a vacuum, is never truly empty – fluctuations in the quantum foam can manifest matter and energy as if from nowhere – dark energy is the result of empty space.

We've also seen how such vacuum energy has an anti-gravitational, or repulsive, force that can inflate space-time.

If we combine the 'matter and energy from nothing' and 'space-time expanding' properties of this sort of 'false vacuum', we might be able to build a multiverse.

1. Start with a seed of false vacuum about a billionth of the size of a proton.

2. The false vacuum energy is gravitationally repulsive – so it inflates a 'bubble' of space-time around it.

3. Because vacuum energy doesn't thin out as space-time expands, its density remains constant. So every time our universe seed doubles in size, so does its energy (and its inflating energy doubles along with it).

4. This energy decays into a churning plasma of fundamental particles (electrons, quarks, etc), which also maintains its density as space expands.

Matter condenses from energy

Young Universe

5. From here, the Universe develops in line with the normal Big Bang model – fundamental particles become the complex particles used to build stars and galaxies.

New universe bubble

6. But here's the twist: the false vacuum doesn't decay evenly and, if there is any 'leftover' false vacuum kicking around (maybe between the pages of this book?), this will form the seed for a new Universe 'bubble' in which some left-over false vacuum will create another Universe and so on and so on ...

SLICED MULTIVERSE LOAF

M-theory (an offshoot of string theory) suggests that our three-dimensional Universe exists on a membrane that can be likened to a slice of bread. On the slice are all the stars and galaxies of our Universe, but parallel to our slice are thousands of other universe slices – arranged in a sort of huge cosmic loaf – that butt up against our own but that we can't detect.

It is thought that this might account for the apparent weakness of gravity (compared to the other fundamental forces). It might be spread out through the whole cosmic loaf – with each slice only experiencing a fraction of the total gravitational force.

It is the same problem we once had with our planet. Looked at in isolation, the Earth seems to have been perfectly 'designed' for the creation of life – just the right distance from just the right sort of star with just the right atmosphere and just the right sort of magnetic field (and so on). Of course we now know that there are countless other planets out there where conditions aren't perfect and where life doesn't exist. We were just the winners in the planetary lottery.

The multiverse would solve the problem of our 'perfect' Universe in the same way. Just as the Earth won the planetary lottery, our Universe won the cosmological lottery. It seems 'perfect' because the conditions within it allowed us to evolve and marvel at its perfection. However, there are countless other universes where conditions just aren't right.

You can compare it to a game of cards. If you were allowed to pull just one card from the deck, the chances of pulling out the card you were looking for is quite small, but, if you were allowed to go through the whole deck, its discovery becomes inevitable. The same applies to the multiverse, with infinite permutations of the laws of physics available it is inevitable that one Universe would be perfect for life.

In many ways, a multiverse is a more comfortable concept to get to grips with than a perfect Universe born from the void. And, when it comes to contemplating our Universe's ultimate demise, it's rather more comforting to think that it won't be dying alone – that somewhere, just beyond our membrane of reality, there might be other life forms trying to answer the question:

'How do you build a Universe?'

THE END
(literally)

GLOSSARY

ANTIMATTER The mirror-image of matter in which electric properties are reversed but other properties are identical. For example, the antiparticle of the negatively-charged electron is the positively-charged positron. When a particle meets its antiparticle, they annihilate and all their mass is converted to energy.

ASTRONOMICAL UNIT (AU) Unit of distance used by astronomers equivalent to the average distance between the Earth and the Sun – about 149 million km.

ATOM From the Ancient Greek word *atomos*, meaning 'indivisible'. The atom is the basic unit of ordinary matter and is made up of a nucleus of protons and neutrons surrounded by a cloud of orbiting electrons.

BARYON A particle made up of three quarks. Protons and neutrons are baryons. Astronomers use the term 'baryonic matter' to distinguish the stuff that makes up stars, planets and you and me from dark matter.

BILLION Equal to a thousand million (1,000,000,000,000). The older definition of equal to a million million, isn't used in this book.

BLACK DWARF A white dwarf star that has cooled so that it no longer emits heat or light. Because it is thought to take longer than the current age of the Universe (13.82 billion years) for a white dwarf to cool, no black dwarfs are thought to currently exist.

BLACK HOLE A region of space where the fabric of the Universe (spacetime) is so distorted by a massive body that not even light can escape its gravity. Most galaxies have a supermassive black hole (with the mass of several million Suns) at their centre.

BLUE GIANT A massive, hot star, containing many times the mass of the Sun.

BLUE SHIFT A shift in radiation emitted by an object moving towards an observer – the wavelength is squashed and so moves towards the shorter, blue part of the electromagnetic spectrum.

BOSON Sometimes called 'force carriers', bosons are particle 'messengers' that mediate interactions between the fundamental forces and matter.

CEPHEID VARIABLE A star that changes in brightness (pulses) over a period of time.

The period of the pulsation is directly related to the star's brightness – making them a powerful tool for determining distance. Also known as a 'standard candle'.

CMB (COSMIC MICROWAVE BACKGROUND) The radiation 'after glow' of the Big Bang. The CMB is made up of the first light, emitted about 380,000 years after the Big Bang following the era of recombination. It is now visible as a faint glow of microwave radiation across the entire sky.

COSMIC INFLATION An extension of Big Bang theory that suggests the Universe underwent a brief period of exponential expansion less than a second after the Big Bang.

DARK ENERGY A hypothetical form of energy that permeates the Universe, making up 68.3 per cent of the Universe's total mass-energy content. Dark energy has an anti-gravity effect that is thought to be accelerating the Universe's rate of expansion.

DARK MATTER A mysterious form of matter that only interacts with 'normal', or baryonic, matter through gravity. Although it can't be seen directly, its existence and properties have been inferred from how its gravity affects visible matter and radiation. Dark matter makes up 26.8 per cent of the Universe's total mass-energy content.

DRUNKARD'S WALK A term used to describe the motion of a photon within dense plasma, such as existed in the early Universe (before recombination) and within stars. In such a dense, high-energy environment, photons are continually absorbed and reemitted by other particles.

ELECTROMAGNETIC FORCE One of the fundamental forces. The electromagnetic force affects any fundamental particle that carries a charge. Its force carrier is the photon.

ELECTROMAGNETIC RADIATION A form of energy that propagates throughout the Universe as electric and magnetic waves. Visible light is just one part of this spectrum that stretches from high-energy, short-wavelength gamma rays, through X-rays, ultra-violet, infrared, visible light and microwaves, all the way to low-energy, short-wavelength radio waves. Usually pictured as a wave, electromagnetic radiation can also be described as a stream of photons.

ELECTRON A negatively-charged elementary particle that orbits the nucleus of an atom.

ELEMENTARY PARTICLE Also known as a fundamental particle. These are particles that, unlike baryons or fermions, are not made up of smaller particles and so cannot be divided. Quarks and electrons are examples of elementary particles.

ENTROPY A measure of the disorder within a system. Basically, it means that an organized system of energy is inherently unstable and will seek stability by becoming disorganized.

FERMION The name given to the family of particles that includes fundamental particles, like quarks and leptons, and composite particles like baryons and mesons, which include protons and neutrons.

FORCE CARRIERS Particles that mediate interactions between matter and the fundamental forces. The photon and gluon are examples of force carriers.

FUNDAMENTAL FORCES The four forces (strong force, weak force, electromagnetism and gravity) that act between bodies of matter. The particles associated with these forces are known as the force carriers.

FUNDAMENTAL PARTICLE (*See* Elementary particle)

GALAXY A gravitationally bound system of dust, gas and stars – ranging between hundreds and hundreds of thousands of light years across and containing many hundreds of millions of hundreds of billions of stars. Most galaxies are thought to contain a supermassive black hole at their centre.

GLUON The force carrier particle (boson) of the strong force.

GRAVITY The weakest of the fundamental forces, but the only force that acts on an astronomical scale. Gravity, as described by Albert Einstein's Theory of General Relativity, is the result of massive objects curving the fabric of spacetime. According to Newtonian physics, gravity is the force of 'attraction' felt by a less massive object as it approaches a more massive object.

HADRON The name given to the family of particles made up of two or more quarks. Baryons are hadrons made of three quarks. Mesons are hadrons made up of two quarks.

LEPTON The family of particles that includes electrons and neutrinos.

LIGHT YEAR Unit of distance used by astronomers equal to the distance travelled by light in a year: 9.4 billion km.

MASS A measure of the amount of matter within an object and how much gravity it exerts. Not to be confused with weight, which is the force exerted on an object by gravity.

MESON A sub-group of the bosons. A particle with two quarks.

MILKY WAY Our home galaxy. It contains 200 billion stars and measures 100,000 light years.

M-THEORY An off-shoot of String Theory that proposes that space is made up of 11 dimensions – adding seven spatial dimensions to the three dimensions of space we are familiar with and time.

MULTIVERSE The proposed existence of a (potentially) infinite number of universes that exist beyond, or parallel to, our own.

NEBULA A cloud of dust and gas (from the Latin word for 'fog', or 'cloud'). There are several categories of nebulae. A planetary nebula, for example, is the cloud of dusty debris thrown out by a dying star. A molecular cloud is a nebula where gas is dense enough to act as a stellar nursery for new stars.

NEUTRINO A lightweight particle produced as byproducts of nuclear fusion.

NEUTRON A particle with no electric charge, made up of three quarks. Along with the proton, forms the nucleus of an atom.

NEUTRON STAR The dense collapsed core of a dead star made almost entirely of neutrons, in which the mass equivalent to the Sun is squeezed into a sphere the size of a large city.

NUCLEAR FUSION Also known as thermonuclear fusion, this is the process that powers stars in which two or more atomic nuclei are forced together to create a single, heavier nucleus. The new nucleus is lighter than the sum of the nuclei that made it, so the 'leftover' mass is liberated as energy.

NUCLEOSYNTHESIS The formation of heavy chemical elements through the fusion of lighter elements in thermonuclear fusion reactions in the cores of stars and supernova explosions.

NUCLEUS The inner core of an atom, made up of protons and neutrons, contains almost all the mass of an atom. The nucleus is orbited by a cloud of electrons.

PHOTON A particle of light; the smallest possible unit of electromagnetic radiation. The photon is carrier of the electromagnetic force.

PLANCK LENGTH The smallest possible unit (or quanta) of length.

PROTON A positively-charged particle made up of three quarks. Along with the neutron, forms the nucleus of an atom.

PULSAR A rapidly rotating neutron star that emits beams of high-energy radiation from its poles.

QUARK A fundamental particle out of which all baryons, such as protons and neutrons, are made. They come in six varieties, known as 'up', 'down', 'strange', 'charm', 'top' and 'bottom'.

QUASAR The very bright active core of a galaxy, fueled by matter being sucked into a supermassive black hole. Quasars can outshine all the stars of the galaxy that surround it – shining brighter than a hundred billion Suns.

RED DWARF A small, cool star.

RED GIANT An aging star that has exhausted its supply of hydrogen and moved on to fusing heavier elements, which heats up the star causing the star to inflate and its surface to cool (and so appear red).

RED SHIFT The reddening of radiation emitted by an object moving away from the observer – the wavelength is stretched and so moves towards the longer, red part of the electromagnetic spectrum.

STANDARD MODEL A unifying theory of physics that describes the interaction between the fundamental forces (excluding gravity) and elementary particles of matter.

STEADY-STATE THEORY A now-debunked rival to Big Bang theory, which described the Universe as being in an eternal, constant state of expansion, with no beginning and no end.

STELLAR NURSERY A region of dense gas where stars are born and grow.

STRONG NUCLEAR FORCE The strongest of fundamental forces but with the shortest range. It holds together to form protons and neutrons. Its force carrier is the gluon.

SUPERMASSIVE A term used to describe an object (usually a black hole) with a mass several million times that of the Sun.

SUPERNOVA The explosive demise of a star. There are two main types of supernovae: Type I and Type II. Type I are caused by the dramatic collapse of a stellar remnant, such as a white dwarf, which ignites a runaway thermonuclear reaction and explodes. Type II are caused by the collapse of the core of a massive star, which unleashes a shockwave of energy that blasts stellar material into space.

UNCERTAINTY PRINCIPLE The principle thought up by Werner Heisenberg that you can never know the exact position and movement of a particle. The more accurately you know one, the less accurately you can know the other.

WAVELENGTH The distance between two crests of a wave.

WEAK NUCLEAR FORCE The second weakest of the fundamental forces. Responsible for radioactive nuclear decay, its force carriers are the W and Z bosons. It has the shortest range of all the fundamental forces.

WHITE DWARF The dense, cooling remains of a star not massive enough for fusion reaction beyond carbon to take place. Most stars, including the Sun, end their days like this.

INDEX

stars 105, 112
matter 36, 38–9, 50
    death of 206
    elementary particles 63
    and energy 69, 110
    lumpy distribution of 47, 90, 91
    and protogalaxies 99
    see also dark matter
Maxwell, James Clerk 57
Mendeleev, Dmitri 54–5
Mercury 182
Messier, Charles 11–12, 13, 26
Milky Way 10, 12–13, 26, 153, 156–9
    age and naming of 141
    and the expanding Universe 206
    supermassive black hole in the 153, 156, 157, 159
Moon 10, 182, 183
    age of 27
    and life on Earth 192
    and parallax movement 20
Mtheory 84, 85, 211
multiverse theory 206–11
    and black holes 207–8, 209
    building a multiverse 209
    false vacuum inflation 208, 210
muons 60, 62

nebula clouds
    and the solar system 162–6, 167
Neptune 170, 185, 188
neutrinos 102–4
neutrons 58–60, 81
    and fundamental forces 72–3, 74, 75
    and quarks 37, 38, 62
    and stars 100, 119, 120–1
neutron stars 84, 113, 123–8
    and black holes 129, 130
    collision of 125–7, 128
    and pulsars 112, 126–7
Newton, Isaac 57, 83, 84–7, 96
nitrogen 110
normal matter 50
nuclear fusion 39, 40, 75
    and protostars 103–4
nucleic acids 191
nucleosynthesis 114–17, 118–23, 127–8

Oort Cloud 189
organic compounds 191–2, 193
oxygen 110, 123, 191
    and planetbuilding 172, 173, 175

parallax measurement 19–22, 23
particles 68
    double slit experiment 64–5

elementary 62–3, 78–9
and energy 69
mass and the Higgs field 78–81
and models of the atom 56–61
and the new Universe 91
Standard Model of particle physics 60–1, 62, 63, 68
virtual particles 66–7, 81
waveparticle duality 65
see also bosons
Pauli, Wolfgang 68
Penguin galaxy 137
Penzias, Arno 46, 47
Periodic Table of the Elements 55, 59
photons 62, 79, 106
    birth of the Universe 36, 37, 38, 39, 43, 44
    and electromagnetic force 76, 77
    and gravity 83, 87
    and molecular hydrogen 98–9
    polarized 35
    the Sun and the 'drunkard's walk' 40, 41, 168
    virtual 76
Pius XII, Pope 17
Planck distance 58
Planck era 33–4, 38
Planck length 50
Planck, Max 50, 58, 68
planetary nebulae 113, 122, 123
planetoids 167, 186
planets
    building 166, 167, 171–89
    gas giants 167, 170, 184–8, 188
    and gravity 83
    ice giants 170, 185, 189
    measuring 20
    protoplanetary disks 166, 167, 171, 174–7
    rocky 171–84
plasma 39, 40–1, 51, 166
    thermonuclear 110
Pluto 170
primeval atom theory 16, 17, 207
probability clouds 61, 68
proteins 191
protogalaxies 91–3, 97–9, 136
    and black holes 142–3, 144
protons 34, 41, 56–7, 69
    and Einstein's gravity 82
    and fundamental forces 72–4, 75, 76, 77
    mass 81
    and neutrons 58–60
    and plasma 39, 40–1
    protonproton cycle 102, 114, 116, 119
    and protostars 100, 101, 102–3
    and quarks 37, 38, 62
protoplanetary disks 166, 167, 171, 174–7

ABOUT THE AUTHOR

Ben Gilliland is an award-winning science writer and illustrator who never really set out to be one. He started his popular MetroCosm science column in the UK's Metro newspaper in 2005 because he was fed up explaining science to journalists and he figured he's have a go. Much to his surprise he now makes a living from it

AUTHOR THANKS

I would like to thank (in no particular order): my parents, Alan and Pauline, for having me and instilling a life-long love of learning; my wife, Charlotte, for her patience, support and general loveliness; my daughter, Jasmine, for making me grow up (but not too much); Jenny Campbell, for giving a feckless Graphics Editor his own science page; Dave Monk, for catching all those erroneous apostrophes; Heather MacRae, for her tireless energy and support.

Commissioning Editor: Hannah Knowles
Senior Editor: Ellie Smith
Editor: Pollyanna Poulter
Art Director: Jonathan Christie
Book Design: The Oak Studio
Senior Production Manager: Peter Hunt